Lead in the World of Ceramics

A Source Book for Scientists,
Engineers, and Students

738.134

Compiled and Edited by
JOHN S. NORDYKE
Vice President
HAMMOND LEAD PRODUCTS, INC.

Sponsored by The International Lead Zinc Research Organization, Inc.
New York, NY   10017

Published by The American Ceramic Society, Inc.
Columbus, OH   43214

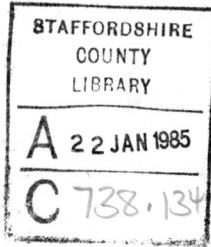

Library of Congress Cataloging in Publication Data

Lead in the world of ceramics.

"Sponsored by the International Lead Zinc Research Organization, Inc."
Includes bibliographical references and index.
1. Ceramics—Congresses. 2. Lead compounds—Congresses. I. Nordyke, John S.
II. International Lead Zinc Research Organization. III. American Ceramic Society,
TP786.L43   1984   666   84-11116
ISBN 0-916094-57-X

Printed in the United States of America.

Lead in the
World of Ceramics

Dedication

For his lifelong devotion to excellence and for his unselfish leadership in the lead industry, this book is respectfully dedicated to William Peter Wilke III.

A Note of Appreciation

We express our appreciation for the work and cooperation of those authors who have contributed text material for this book. Aside from those chapters written by the Editor, eleven chapters have been written by contributing authors. Some of these chapters have been written by one person, others by as many as six. All were selected for their outstanding expertise in the subject covered. All have cooperated beautifully with us.

The following were selected as collaborators:

George B. Hares	Corning Glass Works
Willis H. Barney	Corning Glass Works
Francis W. Martin	Corning Glass Works
Herbert E. Rauscher	Corning Glass Works
Thomas J. Loretz	Applied Fiberoptics, Inc.
Gary W. Ganzala	Owens-Corning Fiberglas, Inc.
John T. Jones	Lenox China, Inc.
Robert C. Ortman	Mayco Colors
Lester M. Dunning	Ferro Corporation
Jack Wells	Owens-Illinois, Inc.
Laurence D. Gill	Pemco Products, Mobay Chemical Corp.
Richard A. Eppler	Pemco Products, Mobay Chemical Corp.
John B. Blum	Rutgers University
Jack Liker	Monogram Ind., Mykro/Mycalex Div.
Donald J. Tucker	Consultant
Neil V. Maresca	International Lead Zinc Research Organization
Malcolm G. McLaren	Rutgers University
Jerome F. Smith	Lead Industries Association, Inc

Their contributions are greatly appreciated.

A special note of appreciation is due to William R. Prindle, Corning Glass Works, for his valuable support of this book.

John S. Nordyke

Foreword

This book is intended to fill a need for a comprehensive review of the properties and uses of lead compounds in the area of ceramics. It is designed to provide engineers, chemists, and ceramists with the latest and most complete information on the use of lead compounds in all types of ceramic products and processes. The ultimate objective is the expanded use of lead chemicals in ceramics in an economically viable and environmentally acceptable manner.

An earlier manual, *Lead in the Ceramic Industries,* was published in 1956 by the Lead Industries Association as a convenient guide for the industrial ceramist. It provided typical formulas, firing procedures, and resulting properties. During the intervening quarter-century, many new applications of lead compounds were developed and commercialized in the ceramic industries. Additionally, ILZRO research contributed to the establishment of international standards for sampling, testing, and permissible limits of lead and cadmium release from ceramic foodware surfaces to assure the safety of the consuming public. These developments, along with results of other ILZRO research conducted over the past 2 decades, are incorporated in this volume.

Because of the diversity of applications for lead compounds in such products as ceramic glazes, glasses, enamels, colors, and electronic ceramics, and the need for plant safety and good occupational health practices, ILZRO called on John S. Nordyke, Vice President of Hammond Lead Products, Inc., to coordinate the preparation of chapters in each of these areas. On the basis of more than 50 years of experience, including the presidency of the American Ceramic Society, Mr. Nordyke was able to assemble a group of experts to write the required chapters in their own areas of expertise. Mr. Nordyke is eminently qualified for this task. Since graduating (with a B.S. in Chemistry) from the Oklahoma A & M College in 1931 (now Oklahoma State University), he worked with Eagle-Picher Lead Co. for 22 years in manufacturing, research, and sales. He has been with Hammond since 1953 in various capacities. During this time, he promoted research and development, which led to several new products, and collected 19 world patents along the way.

ILZRO also expresses its gratitude to the American Ceramic Society for assuming the task of publishing this volume.

Dr. S. F. Radtke
President
International Lead Zinc Research Organization, Inc.

Contents

Introduction

Lead may well have been the first metal to be smelted by early man. Beads of lead found at the site of Çatal Hüyük in present-day Turkey have been dated to about 6500 B.C.[1] (Neolithic). The alchemical symbol for lead, ♭ , is the sign of Saturn, father of the gods. Much of the ancient mining and smelting of lead was carried out to recover silver, which is present in unrefined lead in varying amounts. The so-called silver mines at Laurion in Greece, were in reality, lead mines, with silver as the most desired product. This silver was the main source of the great wealth of Athens during the Age of Pericles, the Golden Age of Greece (fifth century B.C.). Recovery of lead, therefore, must have preceded that of silver. There is strong evidence that smelting of lead preceded that of copper by as much as 3000 years.[2]

In the recovery of silver from lead, the lead was burned at a temperature of about 1100°C, forming a slag of litharge (Greek: silver stone) and leaving unburned molten silver. In the earliest times, the market for litharge (lead monoxide) was small and the excess was cast aside in piles as waste. In later periods, these waste piles became rich sources for the recovery of lead metal.

Pottery is an ancient product, possibly much older than lead. Glass is also a venerable artifact and is one of man's earliest manufactured products. However, lead oxide and pottery did not come together in the form of glazed ware until sometime in the bronze age. Much unglazed pottery of the Neolithic age is displayed in the museums of China, the Middle East, and other areas of the world. William Pulsifer states: "Schliemann found at Mycenae objects resembling buttons which proved upon analysis to be composed of a strongly burned clay varnished with a lead glazing. . . It has also been stated that the glazing of pottery found in ancient Egyptian tombs is composed of lead, and there is also reason to believe that the glazing of bricks and other articles found at Nineveh and Babylon was produced in some cases by the use of litharge. Maigne states that the art was invented in the East and was supposed to have been known in the time of Solomon, and the research of modern archaeologists seems to confirm this suggestion. Eraclius, whose manuscript is attributed to the ninth or tenth century, describes the process minutely and gives directions for producing a variety of colors and tints in the glaze."[2] Thus, it is apparent that the lead and pottery industries have enjoyed a relationship of several thousands of years duration.

Schliemann[2] describes objects of lead glass found at "Spata" (sic) in Greece (Sparta?) thought to date from the eighth century B.C. Eraclius[2] describes the manufacture of lead glass as follows:

"Take good shining lead and put it into a new jar and burn it in the fire until it is reduced to a powder then take it away from the fire to cool; afterward take sand and mix with the powder, but so that two parts may be of lead and the third of sand, and put it into an earthen vase, then do as before directed for making glass."

Beckman[2] tells of an ancient mirror said to have belonged to Virgil, which was accidentally broken. One small piece, when analyzed, "showed that a considerable quantity of lead was used in the manufacture of the glass." "The earliest glass objects so far known date from the middle of the third millenium B.C., at the earliest, and all have been found in western Asia (Mesopotamia, Syria, etc.). They are small, solid objects such as beads, rods, and so forth; the earliest vessels from the same general area are dated on archeological grounds no earlier than the early 16th or 15th century B.C."[3] All of these were opaque, translucent, or strongly colored.

Cristallo, or colorless glass, was developed by the Venetians before the middle of the 15th century, but in reality this was slightly brown or gray tinted. George Ravenscroft of London, by 1676, produced a lead crystal which "in transparency and brilliance came near to crystal."[3] By the end of the 18th century, English lead crystal was the most sought-after glass in the world.[3] One of the three known Ravenscroft goblets is on display at the Corning Museum of Glass.

Enameling of glass and metals is an ancient art form, initially used only for decorative purposes. Brightly colored lead-containing glass, melting at temperatures below that of the vessel to which it was to be applied, was ground to a powder and mixed into a suitable medium, such as an oil. This mixture was then painted onto the surface of the object to be decorated and fired. This procedure, seemingly a reasonable advancement over the earlier process of decoration with threads of brightly colored glass, was used to decorate gold, silver, copper, and bronze jewelry as well as glass. Enameling was well-known in Rome, Egypt, and Syria as early as the first century A.D.

Enameled glassware was manufactured in large quantity in Europe, especially Venice and Germany, from the 16th century onward.[3] Enameled objects of gold, silver, and copper were also very popular. It was not until the early 1800's that enameling of cast iron for bathtubs was developed, with enameling of sheet steel following in about 1850.[4]

Lead, in the form of the oxide, has found an essential place in the manufacture of ceramic products because of many inherent properties essential to these processes and products. The great range of desirable properties, important not only in the melting and forming of glass and the firing of glazes and enamels, but also in the properties of the finished products, will be described in detail in later chapters. Lead oxide is an extremely versatile material that has held its place in ceramics for thousands of years, simply because no other material or combination of materials performs as well.

The proliferation of ceramic products through the centuries has greatly expanded the uses of lead in the form of the monoxide in ceramic products. This is true of the past 100 years, especially since the invention of incandescent and fluorescent lighting, television, the computer, and the host of other electronic inventions and developments, requiring nearly 80 000 tons per year of lead oxide.

This book is presented in an effort to provide, in one cover, as much of today's knowledge of the ceramic applications of lead compounds as is known to us at this time. The ancient background information is essential for perspective, along with the new research, which has grown greatly in volume in recent years. We will not attempt to extrapolate to future years, but we are confident that the future holds much in the way of exciting and invaluable developments, made possible by the properties of this exceedingly versatile material, lead oxide.

Production of lead may rival that of pottery in its antiquity. While those activities are not the oldest profession, the production of lead may well be the oldest industry.

References

[1]N. H. Gale and Z. Stos-Gale, "Lead and Silver in the Ancient Aegean," *Sci. Am.,* **244** [6] 176–7, 181–6, 188, 190–2 (1981).

[2]W. Pulsifer; History of Lead. D. Van Nostrand, New York, 1888.

[3]Robert J. Charleston; Masterpieces of Glass, A World History from the Corning Museum of Glass. Harry N. Abrams, Inc., New York, 1980.

[4]A. L. Friedberg; Encyclopedia of Chemical Technology, Vol. 9, 3rd ed. Wiley, New York, 1980.

1
Lead Compounds and Their Properties

Four general types of lead chemicals are manufactured for ceramic use. These are litharge, red lead, white lead, and the lead silicates. Several variations of each product are prepared so that each may be adapted to a wide variety of uses.

Litharge (Lead Monoxide-PbO)

Litharge is the basic unit in the chemistry of lead in ceramics. Whatever the compound may be, the portion of principal interest is PbO. Litharge, of course, is 100% PbO. It is the simplest of the lead compounds.

Litharge is used in the manufacture of glass, dry process enamels, and in the production of frits for all purposes. It is manufactured by the oxidation of metallic lead by a variety of processes, each resulting in a distinctive variation in physical properties. As a result, litharge is available in a wide range of particle sizes and in two crystal forms.

The yellow or orthorhombic form is stable at temperatures between about 550 and 888°C (1022 and 1630°F), the melting point. It is metastable at normal temperature. The red or tetragonal crystal form results from oxidation of lead at temperatures around the melting point of lead and below the range where red lead (Pb_3O_4) is formed. A red modification of litharge is formed also when molten litharge is allowed to cool slowly. In this case, it forms a flaky form of tetragonal litharge. When fused litharge is chilled rapidly, it crystallizes in the orthorhombic system, but usually as a flaky variety.

Sublimed or fumed litharge is typical of the orthorhombic form. Most mechanical furnace litharge is formed first in the yellow form but is converted partially to the red form by attrition during the milling operation. This partial conversion alters the color from bright yellow to a slightly tan shade.

The crystal form of litharge, so far as is known, is not of great importance in its use in ceramic processes. It is well, however, to recognize these differences because of the frequent questions in this regard. In processes involving aqueous reactions, the yellow form is much more reactive than the red.

Apparent density, particle size, and chemical purity are the important criteria when litharge is being selected for a specific use in a ceramic process. Apparent density and particle size are determined by the method of manufacture. Extreme purity is quite typical of contemporary manufacturing methods. It is the result largely of starting with pure metallic lead and the use of great care in processing, handling, and packing to prevent contamination.

Purity requirements become more demanding with passing time and new developments. While up to 5 ppm of coloring oxide impurity was considered

Table I. Typical Properties of Lead Compounds

Properties	Litharge — Sublimed or fumed	Milled	Screened	Granular	Red lead	White lead	Lead monosilicate	Lead bisilicate	Lead bisilicate	Tribasic lead silicate
Composition	PbO	PbO	PbO	PbO	Pb_3O_4	$2PbCO_3 \cdot Pb(OH)_2$	$PbO \cdot 0.67SiO_2$	$PbO \cdot 0.03Al_2O_3 \cdot 1.95SiO_2$	$PbO \cdot 0.254Al_2O_3 \cdot 1.910SiO_2$	$PbO \cdot 0.33SiO_2$
Apparent molecular weight	223.21	223.21	223.21	223.21	685.63	775.67	263.27	343.47	363.82	243.27
Equivalent weight	223.21	223.21	223.21	223.21	228.54	258.56	263.27	343.47	363.82	243.27
Specific gravity	9.65	9.55	9.55	9.4–9.5	8.95–9.10	6.8	6.50–6.65	4.60–4.65	4.30–4.35	7.52
Apparent density: Scott volumeter $(g/in.)^3$	8–12	30–40	50–80	70–80	20–25	7–9	200–230	170–200	160–190	230–260
Packing weight $(lb./ft.)^3$	90–100	240–260	280–350	280–350	210–230	90–100	1250–1350	1450–1500	1425–1475	1250–1350
Melting range: °F	1630	1630	1630	1630	1020 decomposes	730 decomposes	1250–1350	1450–1500	1425–1475	1250–1350
°C	888	888	888	888	550 decomposes	390 decomposes	677–732	788–816	774–802	677–732
Coefficient of linear thermal expansion (50–450°C)							9.3×10^{-6}	7.1×10^{-6}	5.9×10^{-6}	
Color	yellow	yellow to tan	yellow	reddish yellow	orange-red	white	light yellow	light yellow	light yellow	reddish yellow
Refractive index	2.67						2.00–2.02	1.72–1.74		
Solubility in water	slight 99%	slight 99%	slight varies	slight usually	insoluble 99%	slight 99%	insoluble 95–99%	insoluble 95%	insoluble 95%	insoluble 95%
Screen analysis	–325 mesh	–325 mesh	–10 mesh	–10 mesh	–325 mesh	–325 mesh	–10 mesh granular; 99% –200 mesh ground; 95% –325 mesh ground	–10 mesh granular; 95% –200 mesh ground	–10 mesh granular; 95% –200 mesh ground	–10 mesh

2

quite good only a few years ago, some new developments now require purity levels measured in a low number of parts per billion. Needless to say, such purity levels are attained at high cost, far removed from the cost of volume commercial lead products.

Red Lead (Pb₃O₄)

Red lead is manufactured by further oxidation of litharge. In this process litharge is roasted at a temperature near 482°C (900°F). The oxidation of the litharge particles proceeds from the surface toward the center and reaches a degree of completion within a given time which is largely dependent on the particle size of the litharge used. Ceramic grades of red lead by choice are usually quite dense with relatively large particle size. The true Pb_3O_4 content is about 75%, the balance being PbO.

Among the merits of red lead is the fact that the extra oxygen that it contains is released at about 550°C (1020°F). Thus, red lead provides an important measure of protection against reduction during the early stages of melting of lead glasses.

Lead Silicates

Probably the most versatile of lead compounds are the lead silicates. They are manufactured in granular, dustless, free-flowing form for use in glass, dry process enamels, and in frit batches, or in finely divided preground form for use in glazes or in certain ceramic bodies. They are, therefore, adaptable for almost any conceivable use by the ceramic industries.

Three types of silicates are available commercially. These are considered separately.

Lead Monosilicate (PbO · 0.67SiO₂)

This composition contains 85% PbO and 15% SiO_2. It is approximately the eutectic mixture of lead orthosilicate ($2PbO \cdot SiO_2$) and lead metasilicate ($PbO \cdot SiO_2$). The name is therefore not accurately descriptive of the composition. In finely powdered form, it is readily soluble in dilute hydrochloric or acetic acid.

Lead Bisilicate (PbO · 0.03Al₂O₃ · 1.95SiO₂)

This material, also known as lead aluminum bisilicate, contains 65% PbO, 1% Al_2O_3, and 34% SiO_2. While its principal use at this time is in pottery and wall tile glazes, a new and growing use is developing in low-temperature bodies and in some of the ultralow-loss dielectric bodies. Lead bisilicate is also used in the manufacture of reflective spheres for highway marking and sign use.

Other lead aluminum bisilicates are also available from frit manufacturers, of which the cone deformation eutectic composition is an example. This composition is $PbO \cdot 0.254Al_2O_3 \cdot 1.910SiO_2$, containing 61.35% PbO, 7.12% Al_2O_3, and 31.53% SiO_2.

Lead bisilicates are extremely resistant to leaching by dilute acids, including gastric juice. For this reason, this type of composition greatly reduces toxic properties and offers the maximum of safety of any of the lead products.

3

Tribasic Lead Silicate (PbO · 0.33SiO₂)

This highly basic lead silicate is manufactured only in granular form and is intended only for use in lead glass where bulk shipping, storing, and handling methods are used. It contains 92% PbO and 8% SiO₂ and is the eutectic of tetralead silicate (4PbO · SiO₂) and lead orthosilicate (2PbO · SiO₂). The powdered form is readily soluble in dilute acids.

In general, the lead silicates are manufactured in highly pure form. The process in use is that of fusion of a mixture of very pure litharge and selected glass sand. Kaolin or pyrophyllite may be used as the source of Al₂O₃ in lead aluminum bisilicates. Mixed batch is usually fed continuously into a melting furnace. The fused lead silicate is usually quenched in water and dried. It may be ground before packing in bags or it may be shipped in granular form, either packed in bags or shipped in bulk in covered hopper cars, bulk trucks, or in portable bins, cartons, or "sling bins" holding 2000–3000 lbs (907–1361 kg).

White Lead (2PbCO₃ · Pb(OH)₂)

Basic carbonate white lead is manufactured by chemical precipitation from a slurry of litharge in water containing a small amount of acetic acid. Carbon dioxide gas is fed in carefully controlled amount and rate into a closed vessel under agitation. The resulting product is very pure, clean, white, and finely divided. With median particle radius of 0.35 μm, it has excellent suspension qualities. The particles are sufficiently fine to be kept in constant agitation by Brownian movement in spite of the specific gravity of 6.9. See Fig. 1 for thermogram for basic carbonate white lead and decomposition products.

While basic carbonate white lead has a history of use in ceramics, especially in glazes, spanning many centuries, its use has been greatly diminished in recent years. White lead is readily soluble in many acid solutions, including gastric juice. In contrast, lead bisilicate and some of the pottery glaze frits are relatively insoluble and, therefore, present substantially less hazard to those who handle glaze materials in the plant. There remain, however, special circumstances where white lead is preferred and special precautions are taken to assure safety.

Lead Isotopes

Studies of the ratios of the four stable isotopes of lead (204, 206, 207, 208) present in lead ores from different mining areas has been helpful in the study of archaeological artifacts which contain lead. The distinctive patterns of isotopes in lead mined in various locations has made it possible, in many cases, to determine the sources of lead in ancient glass, glazes, and ancient metallic artifacts.

Robert H. Brill, Administrator of Scientific Research, Corning Museum of Glass, and associates have studied isotope ratios of lead ores and metals from many sources about the world. Similar studies of lead present in ancient artifacts, including metal goods, glass, glazed pottery, and tile have yielded extremely interesting information.

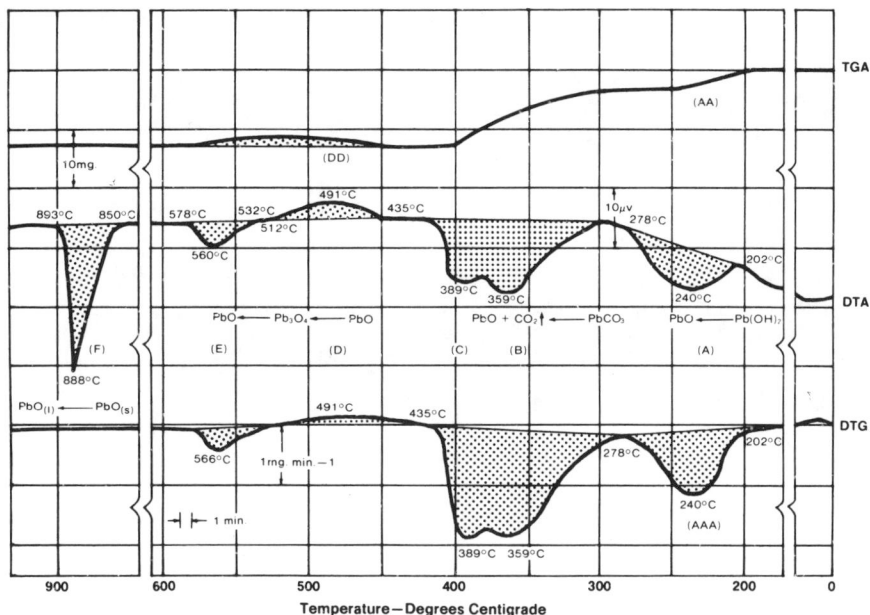

Fig. 1. Thermogravimetric analysis and differential thermal analysis diagrams for basic carbonate white lead and decomposition products. The DTA endotherm in the range 202 to 278°C (A) indicates the change from hydrous basic lead carbonate to the anhydrous form. This reaction may be represented as $2PbCO_3 \cdot Pb(OH)_2 \longrightarrow 2PbCO_3 \cdot PbO + H_2O \uparrow$. The loss of weight and rate are shown on the TGA and DTG plots (AA) and (AAA). At 359 and 389°C, (B) and (C), there are two endotherms, with accompanying loss of weight indicating a two-stage decomposition of anhydrous basic lead carbonate to lead monoxide in the tetragonal state. This might be represented as $2PbCO_3 \cdot PbO \longrightarrow PbCO_3 \cdot 2PbO + CO_2 \uparrow \longrightarrow 3PbO + CO_2 \uparrow$. Upon continued heating of the material, the DTA exotherm to 435 to 532°C (D) and thermogravimetric weight gain (DD), show the oxidation of lead monoxide to red lead. There follows the decomposition of the red lead to lead monoxide in the orthorhombic state (yellow), as evidenced by the endotherm at 532 to 578°C (E). Lead monoxide melts at 888°C (F), as indicated by the endotherm peak.

We feel that reference to this work and to the writings of R. H. Brill et al. is of value to some of our readers.

Robert H. Brill, "Ancient Glass," Sci. Am., **209** [5] 120–31 (1963).
Robert H. Brill and J. M. Wampler, "Isotope Ratios in Archaeological Objects of Lead"; pp. 155–56 in Application of Science in Examination of Works of Art. Museum of Fine Arts, Boston, MA, 1965.
Robert H. Brill and J. M. Wampler, "Isotope Studies of Ancient Lead," Am. J. Archaeol., **71**, 63–77 (1967).
Robert H. Brill, "A Great Glass Slab from Ancient Galilee," Archaeology, **20** [2] (1967) 88–95.
Robert H. Brill, "The Scientific Investigation of Ancient Glasses"; pp. 47–68 in Proceedings of the Eighth International Congress on Glass, London. The Society of Glass Technology, Sheffield, England, 1968.
Robert H. Brill, "Lead Isotopes in Ancient Glass"; pp. 255–61 in Proceedings of the Fourth International Congress on Glass, Ravenne-Venise. Secretariat of the International Congress on Glass, Liege, 1969.

Robert H. Brill, "Lead and Oxygen Isotopes in Ancient Objects"; pp. 143–64 in The Impact of the Natural Sciences on Archaeology. The British Academy, London, 1970.

Robert H. Brill, "New Directions in Lead Isotope Research"; pp. 73–83 in Application of Science in Examination of Works of Art. Boston Museum of Fine Arts, Boston, MA, 1970.

Robert H. Brill, pp. 279–303 in Lead Isotopes in Ancient Coins. *Spec. Publ., Roy. Numismatic Soc.* 1972, No. 8.

Robert H. Brill, I. L. Barnes, W. R. Shields, and T. J. Murphy, "Isotopic Analysis of Laurion Lead Ores"; pp. 1–10 in Archaeological Chemistry, Chapter 1. The American Chemical Society, Washington, D.C., 1974.

Robert H. Brill, I. L. Barnes, and B. Adams, "Lead Isotopes in Some Ancient Egyptian Objects"; pp. 9–27 in Recent Advances in Science and Technology of Materials, 3. Plenum, New York and London, 1974.

"Some Miniature Glass Plaques from Fort Shalmaneser, Nimrud: Part II, Laboratory Studies." *Iraq,* **XL** [Spring] 23–29 (1978).

Robert H. Brill, Kazuo Yamasaki, I. Lynus Barnes, K. J. R. Rosman, and Migdalia Diaz. "Lead Isotopes in Some Japanese and Chinese Glasses." *Ars Orientalis,* **XI**, 87–109 (1979).

Robert H. Brill, I. L. Barnes, J. S. Gramlich, and M. G. Diaz, "The Possible Change of Lead Isotope Ratios in the Manufacture of Pigments: A Fractionation Experiment"; in Archaeological Chemistry II. The American Chemical Society, Washington, D.C., 1978.

2
Properties Imparted by
Lead Compounds to Ceramic Products

Lead is one of the most versatile of all raw materials used in the ceramic industry. It is employed in a great variety of ceramic products for a great number and combinations of reasons. When we use the word "lead" in its general form, we are referring, of course, to the oxides, the silicates, and the basic carbonate of lead, never to the metallic form. In each of these compounds there is one common denominator, PbO, or lead monoxide. It is this portion of the various lead compounds that is significant in the ceramic applications of lead products.

All uses of lead chemicals in the ceramic industries are related to the glass-modifying or forming properties of lead monoxide. A glassy coating on a whiteware ceramic body, which we would call a glaze, a coating of glass on metal, which we would call an enamel, or an object composed entirely of glass, all present the basic condition of the glassy state of matter. The use of lead compounds in low-temperature vitrified ceramic bodies is no exception, since the attainment of the glassy state is a necessary condition in this use as well. PbO is one of very few raw materials which can be used in almost unlimited proportions in glass. It is one of the classic network modifiers in glass formation.

In describing the properties of lead glasses possessed by virtue of their lead content, we shall attempt to show, in a general way, the significance of their basic properties in the various phases of ceramics. The properties of the several lead products available and their relative merits will be discussed in later chapters.

First of all, lead oxide is a powerful flux. As such it aids in the melting and solution of other batch materials. Many other materials are well-known for their fluxing power and, generally speaking, are freely available. In many cases they are less costly than lead compounds. The particular values of lead then must lie in its versatility or the variety of properties of glass possessed because of its lead content. Let us consider some of the more important of these. In so doing, we must consider the working properties, or the effect of the presence of lead in the glass on its melting and working qualities, as well as the qualities possessed by the finished objects themselves.

Low Melting Range

Generally speaking, lead glasses melt at relatively low temperatures. Within limits, the higher the lead content, the lower the melting temperature. This, of course, depends on the total composition of the glass, glaze, or enamel.

7

Low-temperature solder glasses, enamels for aluminum and low-temperature enamels for steel, and glazes for low-temperature ceramic bodies depend on a high lead oxide content for the combination of low firing temperature and durability which they provide.

Wide Softening Range

The successful blowing of fine art glass requires an extremely high order of skill. It also requires, among other things, a glass that can be blown and shaped over a wide range of temperatures. There are glasses that pass very quickly from a highly fluid state to an extremely viscous state over a relatively narrow temperature range. These glasses, even if otherwise acceptable, would present an almost insurmountable problem to the glass blower. Lead glasses provide a wide range of workability, thus facilitating what at best is a difficult art.

Fabrication of incandescent bulbs and fluorescent lighting, along with sign-light tubing, requires a high degree of adaptability. The wide softening range, along with low softening temperature, good electrical properties, and other desirable qualities, makes glasses of high lead content ideally suited for these uses.

The quality of wide softening range has great value in the firing and maturing of glazes and enamels as well. A temperature variation of 15–30°C through the cross section of a kiln is not at all uncommon. Lead glazes usually can be depended upon to mature properly in spite of such variation. Occasional errors in batching may also be overcome by the foolproof nature of lead glazes and enamels. The economic importance of this property can hardly be overestimated when we consider the cost of bringing a piece of dinnerware or other ceramic object to the point where the glaze or enamel is to be applied and fired. Losses due to improper firing are tremendously expensive. While other types of glazes or enamels can be formulated without lead, an attempt to save a little on raw material cost can be far more than offset by the cost of loss due to underfired or overfired ware.

Low Surface Tension

A general characteristic given to glasses, glazes, and enamels by their lead content is low surface tension. This property is of rather great importance in many ways and presents a marked contrast to the high surface tension characteristic of some recently developed leadless compositions.

Low surface tension is the property that is chiefly responsible for the smooth flow and generally high gloss of lead glasses, glazes, and enamels. Low surface tension coupled with the wide softening range of lead glazes accounts for their superior maturing qualities. Low surface tension and low interfacial tension go hand in hand. Low interfacial tension contributes greatly to good wetting of a surface by any liquid, including molten glass, glaze, or enamel, thus promoting good adherence.

8

Low surface tension is the one property that contributes most strongly to the ability of lead glazes and enamels to heal over blisters, drying cracks, and other accidental surface imperfections during firing. The dollar saving by reduction of losses from these causes is a matter of great economic importance.

Resistance to Devitrification

The presence of a moderate-to-high lead content in a glass provides protection against surface crystallization. This tendency toward devitrification can be most annoying and costly when a glass must be heated and reheated in the process of fabrication and forming.

High Index of Refraction

The brilliant fire of a fine piece of crystal glass and the brilliance of glaze on a piece of fine china pay tribute to their high lead content. Other fluxes may be used in glazes and in glass, but the product will be dull and lifeless. None can compare with lead in the brilliance and beauty that it imparts. High index of refraction, variable and controllable with lead content, accounts for the extensive use of lead flint glass for optical and ophthalmic purposes.

Freedom from Color

Lead oxide in itself is not a coloring material except when a very high percentage is used. With the extremely pure materials available today, concentrations of over 60% may be used before a faint yellowish color begins to appear.

Resonance

The clear, resonant, bell-like tone that can be produced by striking a lead glass goblet is a unique property imparted to glass by a moderate-to-high lead content. No other material can give this property to glass. For this reason, it has become a hallmark of quality in fine glass.

High Opacity to Radioactive Energy

The ability of a material to absorb gamma radiation is largely a function of its density. The value of lead metal or lead glass as a shielding material results from this property of high density. The particular added advantage of lead glass comes from its transparency, which permits its use in radiation protective windows through which dangerous operations may be observed. Lead content for radiation absorptive glass ranges as high as 81%.

Good Electrical Properties

Very important properties of glasses, glazes, enamels, and insulator bodies with a significant lead content are high surface and volume electrical resistivity, low power loss, and high dielectric strength. Some of the largest users of lead products in the ceramic industries depend on the electrical properties provided by a moderately high lead content.

Lead glass for neon sign tubing, for support of filaments in incandescent and fluorescent lighting, and for television picture tubes and a host of other electrical and electronic applications, usually contains about 30% PbO. Other formulas are used containing somewhat lesser quantities for a great variety of electrical applications.

Many types of electrical resistors are coated with a glaze of high lead content. An example of such an application where high surface resistivity is essential is the spark plug glaze. Many dielectric bodies achieve their high volume resistivity and low power loss in large part by their high lead content.

Adaptable Coefficient of Expansion

It has been customary to refer to "high coefficient of expansion" as a general property contributed by lead to glass. This is not strictly accurate since there are some notable exceptions. It might perhaps be more accurate to refer to "a wide expansion range" or to "adaptable coefficient of expansion." In most cases a relatively high coefficient is imparted to glass by its lead content, which contributes to the ability of lead glasses and enamels to fit the expansion of metals to which they are sealed or fused. On the opposite extreme are the ultralow-loss dielectric bodies of high PbO content, some of which exhibit low thermal expansion for this type body and possess zero firing shrinkage.

This seemingly paradoxical situation might be explained by a related property of lead glasses, glazes, and enamels. This property is that of toughness or resiliency. A. V. Bleininger described it by saying: "Lead puts rubber into a glaze." This expression may explain in a way the unique ability of lead glasses to conform to such a wide range of expansion requirements.

Summary

The foregoing are some of the more important characteristics of lead glasses derived from their lead oxide content. The decision to use a lead glass for a given purpose will, as a rule, depend on a combination of several of these and seldom on one alone.

The wide variety of properties discussed should serve to emphasize the great versatility of lead compounds in a vast assortment of ceramic formulations and will help to explain why lead can do so many things so well.

3
Lead Glasses

As mentioned in the Introduction, lead (or lead oxide) has been used as a constituent in glass formulations for over 3000 years. Opaque yellow glasses containing lead antimonate ($Pb_2Sb_2O_7$) as the opacifying agent were produced as early as 1450 B.C.[1] Turner reports analyses of a number of ancient glasses containing lead.[2] Among these are a blue lapis lazuli glass from Nippur, Mesopotamia, dated about 1400 B.C., containing 15% PbO; an Assyrian sealing wax red, dated 500–700 B.C., containing 23% PbO; as well as some interesting Chinese glass beads from about 200 B.C. containing 43% PbO and 13% BaO.

Early Russian glasses containing lead are reported by Besborodov.[3] The chemical analyses of some of these glasses are shown in Table I. These ancient Russian glasses belong to two main types of lead glasses: (1) lead oxide–silica, and (2) potassium oxide–lead oxide–silica.

In the first type, lead oxide and silica comprise about 94% of the total weight of the glass. A glass formulation of these proportions could arise from using one part of sand to two parts of lead. The latter would probably be calcined to the oxide before adding to the sand.

This glass would melt easily at low temperatures (800–1000°C) and would be an excellent flux for the various coloring oxides.

The second type of glass is more viscous and would require a higher melting temperature (1100–1200°C). The general composition is very close to that of modern lead crystal. It would have a long working range and would not be prone to devitrification upon repeated reheating, making it especially suitable for the production of hollow ware and windows.

Potash would be obtained by leaching wood ashes, filtering the solution, and allowing it to evaporate to dryness. The raw material ratio would be one part of potash, one part of lead, and two parts of sand. Again, the glass would be a good flux for the various coloring constituents.

The sand was not pure silica but often contained iron oxide, aluminum oxide, lime, and magnesia. The potash probably contained some sulfates. Notice the presence of SO_3 in the K_2O–PbO–SiO_2 glasses.

While this may be the earliest reported use of potash–lead glasses as articles of commerce, the origins of the lead crystal industry in the Western world lie in the work of George Ravenscroft about 1675.

Previous to this date, the glassmaking center was in Italy where soda–lime compositions were used in the production of glass both for utilitarian and luxury purposes. The traditions of glassmaking, carried by Italian craftsmen, spread throughout Europe like wildfire in the 16th and 17th centuries. The Venetian tradition was established in England about 1575 and remained fairly active

Table I. Chemical Analyses of Some Ancient Russian Glasses of the 11th–13th Centuries A.D.

Place Type of ware	Kiev Mosaic	Kalinin-Skaja Bead	Kalinin-Skaja Bead	Smolensk Bracelet	Kiev Bracelet	Kiev Window	Kiev Hollow ware vessel	Kiev Goblet
Color	yellow	reddish brown	violet	green	dark violet		green	colorless
Analysis No.	74	21	25	65	215	111	155	192
SiO_2	32.60	33.68	56.12	31.67	48.67	55.70	30.80	57.97
SnO_2	3.32							
TiO_2	Trace	1.72	0.99	1.41	1.69			
Al_2O_3	0.18					2.20	0.41	1.88
Fe_2O_3	0.12	1.64	0.31	0.85	0.17	0.12	0.07	0.12
CaO	0.20	0.25	0.42	0.20	1.88	2.88	1.08	2.30
MgO	Trace	0.13	0.16	0.36	0.28	0.58	0.07	0.77
PbO	63.01	60.36	23.27	63.46	26.28	23.33	63.82	21.93
CuO							0.60	
Mn_2O_3			1.96		1.50			
Ca_2O_3		0.63						
SO_3			0.53	1.02	0.86	0.47		0.70
Na_2O	0.69	0.73	0.93	0.83	1.35	2.17	0.94	0.82
K_2O		0.51	14.44	Nil	11.40	13.13	1.58	13.36

through the Puritan period and into the Restoration (1660). At this time in England, a very important change in outlook occurred that led eventually to the development of the English lead crystal, variations of which are still being used in fine glassware even to this day. Not only was there a spirit of scientific inquiry and research (the Royal Society was founded in 1662), but there was also a great effort toward national self-sufficiency. Most of the glassmaking ingredients were being imported.

George Ravenscroft was engaged by the Glass-Sellers Co. in 1673 to begin research to eliminate the need of foreign ingredients. A decade before (1662), Christopher Merrett, a Founder Fellow of the Royal Society, had translated into English an important book on glass technology by Antonio Neri, *L'Arte Vetraria,* originally published in 1612. Several chapters were concerned with glass to be used for gemstones; this glass was produced by introducing of "calcined lead" as a flux for sand. Whether or not Merrett's translation influenced Ravenscroft is a matter of speculation.

For a supply of silica, Ravenscroft used English flints, a very hard and pure form of quartz. These were pulverized before mixing with alkali, which was potash rather than the sodium-containing ash used by the Venetians and previously imported by the English from Spain. The combination of high potash and low lime (due to the purity of the flints) led to a glass composition that was not stable to atmospheric moisture and would develop a "crizzling" or network of fine surface cracks which obscured the quality of the glass (Fig. 1). This deterioration of the surface was often called "glass disease."

In an attempt to correct this problem, "calcined lead" was substituted for some of the alkali. At first, the quantity was small, but eventually it was increased to as much as 30% by weight of lead oxide. Eventually, a lead glass composition was found that was not particularly susceptible to "crizzling," and Ravenscroft was permitted to use a raven's head seal as his device on glassware to distinguish the new composition from the older, less stable glasses (Fig. 2).

The glass had unusual clarity for the period and because the use of flint pebbles were, in part, responsible for this clarity (being low in iron), these lead glasses were called flint glasses. These flint glasses were of greater brilliance than hitherto known, and they were readily decorated by cutting, being considerably easier to grind and polish than the existing soda–lime–silica glasses. Improvements in Ravenscroft's formulation led to English crystal glass and later, to distinguish from compositions of lower lead oxide content, English full crystal, which had an approximate composition of 55% SiO_2, 33% PbO, and 12% K_2O.[4]

Variations of the lead crystal formula were evident during the 19th century. Analysis of some lead crystals according to Biser[5] are listed in Table II.

The values for Fe_2O_3 look suspiciously high since the total of $Al_2O_3 + Fe_2O_3$ is often reported as R_2O_3 on analysis sheets. Be that as it may, it is interesting to note that the American crystal contained almost a 50–50 mixture of Na_2O and K_2O as the alkalis.

Other crystal variations were the "half-crystals" containing about 15% PbO and, indeed, some so-called crystals with less than 10% PbO.

Fig. 1: Example of crizzling.

Fig. 2. Ravenscroft goblet showing raven's head seal.

Table II. Analysis (%) of Lead Glass

Glass	SiO$_2$	Na$_2$O	K$_2$O	CaO	PbO	Fe$_2$O$_3$	Al$_2$O$_3$
English crystal	51.93		13.67		33.28		
	59.20		9.00		28.20	0.40	
French crystal	48.10		12.50	0.60	38.00	0.50	
	50.18		11.62		38.11	1.30	1.30
American crystal	53.98	6.71	7.60		29.78	1.93	1.93
	54.12	5.58	7.98		31.27	1.05	1.05

The improvement in lead crystal in the 20th century has not been so much a change in the basic glass formulation as it has the marked improvement in the purity of the raw materials, improved refractories, and improved melting furnaces. These improvements have brought about the degree of perfection that can be found in modern lead crystal glasses such as Steuben art glass.

Optical Glass

Historically, there were two basic processes for making window glass by hand, the cylinder or broad glass process and the crown glass process. In the cylinder process, a long cylindrical shape bubble is blown, and the end is heated and opened to form a cylinder. The entire piece is then reheated, split lengthwise, opened out, and allowed to sag into a flat sheet.

The crown process, on the other hand, consists of first blowing a large spherical bubble of glass. A pontil rod is attached to the surface of the sphere at a point opposite to the end of the blowpipe. The blowpipe is cracked off, the end of the sphere reheated, and opened. Next, the entire sphere is reheated, and when the glass begins to be very fluid, the blowpipe and sphere are rotated very fast. Centrifugal force opens the sphere into a fairly flat disk. The circular sheet is next cracked off from the iron and annealed. Earliest disks were perhaps 8 in. (20 cm) in diameter. A little later they were around 2 ft (0.61 m) in diameter. In 19th-century England, sheets were made by this process that were as much as 5 ft (1.5 m) in diameter.

There are three characteristics of glass made by the crown process. The surface was more brilliant than that of cylinder glass, since the surface had not come into contact with any refractory material, as would occur during the sagging operation in the cylinder process. This fire-polished surface was much more desirable for glazing. A second characteristic was that the thickness of the glass decreased with increasing distance from the center. Large panes of this glass were difficult to put into window frames because of this taper.

The third, and perhaps most important, characteristic was a large mass of clear glass in the center of the disk where it was attached to the pontil iron. This mass of glass is known as a bull's-eye or crown. Usually this crown portion of the glass was thrown away or used for glazing windows over doorways or other areas where there was need only for light and not for visual clarity.

The bull's-eyes themselves were shaped somewhat like lenses. An old Venetian book by Garzoni (1585) tells about the *occhiolaxi,* the makers of spectacles who, with marvelous rapidity, fit bull's-eye glass made at Murano into frames of lead. In general, crown glass was made from soda–lime–silica compositions.

Lenses are the very heart of any optical system and have been known since antiquity. Originally, they were probably used as burning glasses. The word "focus," in Latin, meant a hearth or burning place. The modern French word "foyer" is used for both "hearth" and "focus."

The 17th century was one of great activity in the science of astronomy with glass lenses being the key to progress. Galileo built his first telescope in 1609. Larger and larger telescopes were quickly built to give higher magnifications. However, it was difficult to obtain sharp images since it seemed impossible to eliminate the color fringes at the edge of the image. Much effort was expended by the opticians in grinding lenses of different curvatures to eliminate this problem.

In 1666, Sir Isaac Newton discovered that when a beam of sunlight passed through a glass prism the beam would bend and form the different colors of the spectrum. The violet end of the spectrum was more remote from the edge of the prism than the red. This bending of light is called refraction, and the amount the light of any particular color that is bent is measured by the index of refraction. The refractive index of glass for violet light is higher than that for red.

Newton overhastily concluded from his experiments "that all refracting substances diverged the prismatic colors in a constant proportion to their mean refraction," and he drew the natural conclusion that "refraction could not be produced without color" and therefore "that no improvement could be expected from the refracting telescope." Accordingly, he turned his attention to the development of the reflecting telescope.

However, Newton was not correct in his assumption. Chester Moor Hall, of Essex, argued that the different humors of the human eye so refract rays of light as to produce an image on the retina which is free of color distortions, and he reasonably argued that it might be possible to produce a similar result by combining lenses of two different substances. After some time he discovered that a convex lens of crown glass (soda–lime–silica glass) cemented to a concave lens of flint glass (alkali–lead–silica glass) was an answer to chromatic aberration, but one which depended on the improvement of the quality of flint glasses as then known. Flint (crystal) glassmakers were apathetic about improving quality because there were oppressive excise regulations in force and only a small demand for such glasses was anticipated. Hall did succeed, though, in 1733 in making several telescopes that were achromatic.[6]

The use of crown and flint glasses soon became universally accepted. (An 1810 catalog by Jean Chevalier lists optical prisms in both flint and ordinary crown glass.) However, the quality of the glass was not good enough for telescopes requiring large lenses. The glass defects that were a problem were of three types: stones, seeds, and stria. What the glassmaker calls stones are solid inclusions and may be unmelted batch, pieces of the refractory container, or crystals formed by the crystallization (devitrification) of the molten glass during cooling. Seeds are small bubbles in the glass that arise from gases entrapped or given off during the melting process. Large seeds are often called blisters. Stria are caused by differences in the composition of the glass within a given piece. This composition difference may be caused by preferential volatilization, poor mixing of batch materials, segregation of batch materials during melting due to density differences, or even a partial dissolution of the refractory container. Quite often these stria take the form of a string or cord running through the glass piece.

The flint glasses were particularly difficult to melt to good quality. It was such a vexing problem that late in the eighteenth century the French Academy of Science offered prizes for perfect disks of optical flint glass, but in vain. After half a century of research, Pierre-Louis Guinand (a Swiss watchcase maker) was the only man who had succeeded in making large specimens of flint glass, and even he had never succeeded in making an 18-in. (48 cm) diameter disk.[7]

He did make one 18-in. (48 cm) disk after much trouble and exertion, but while it was in the annealing oven, fire broke out. While extinguishing the flames, water found its way into the annealing oven and destroyed the contents.

In principle, his method of obtaining homogenous glass was to stir the glass after the batch had melted down. Guinand developed a hollow cylindrical rod of porous burnt fireclay, which could be suspended in a covered pot at the end of a long mechanical arm (see Fig. 3). By manipulating this rod, he was able to stir the glass and obtain fairly homogeneous glass in some sections of the pot. The glass and pot were allowed to cool, after which the block of glass could be cut and remolded into lens shapes. When the glass was cut, however, sections of good-quality glass and poor-quality glass might end up in the same lens. Guinand's flint glass was 45% silica, 12% potash, and 43% lead oxide.[8]

By accident (his workmen let a large chunk of glass fall off a wheelbarrow), he discovered a method of fracturing the glass melt during cooling in such a way that the homogeneity of each piece was greatly improved. His process, or variations thereof, for making optical glass was used until the mid-1940's, when the continuous melting of optical glass was commercially developed.[9]

In detail though, his process was long a profound secret. Guinand discovered the process about 1790. His son, Henri Guinand, became associated with George Bontemps, who was a director of a glass factory. The partnership did not last. Bontemps, in 1848, went to England and became associated with Chance Brothers while Guinand continued the manufacture of optical glass in Paris. The optical glass firm Guinand founded eventually became the great French optical glass company of Parra-Mantois. As a result, until 1886, there were only two workshops that could make first-rate optical-quality flint glass, Parra-Mantois in Paris and Chance Brothers in Birmingham.

Bontemps, at Chance Brothers, immediately started to produce a soft crown and light flint for camera lenses and a hard crown and dense flint for telescopes.[10]

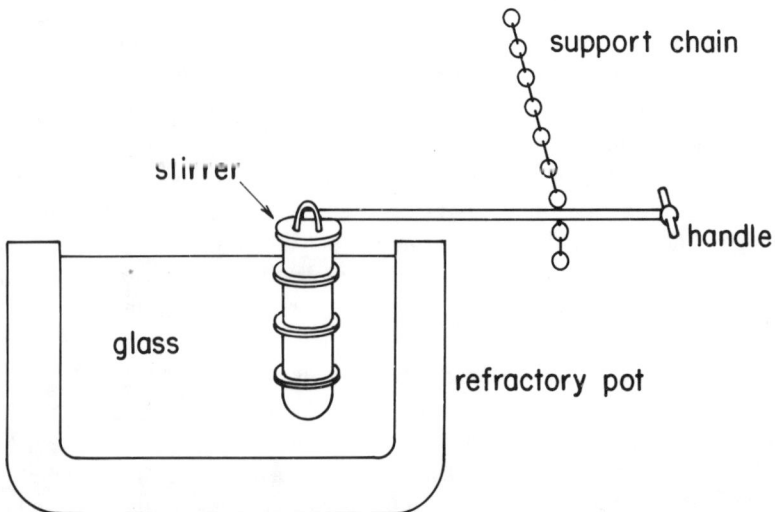

Fig. 3. Model of Guinand's optical glass furnace.

The first dense flint contained equal parts of sand and red lead, but Bontemps kept increasing the lead content until in 1867, he introduced double extra-dense flint, in which nine parts of red lead were used to five parts of sand. Eventually, six types of flint glass were manufactured by Chance Brothers.

Optical Flint Glasses

To understand the terminology of optical flint glasses, a review of optical glass properties is presented.

When light passes from air to a denser isotropic substance such as glass, its velocity is diminished. The ratio between these two velocities is known as the refractive index (n) of the glass.

$$n = \frac{\text{velocity of light in air}}{\text{velocity of light in glass}}$$

If a monochromatic light beam strikes a glass surface at an angle other than the perpendicular, the beam in the glass will be bent toward the perpendicular. The refractive index may be determined from knowledge of the two angles.

$$n = \frac{\text{sine of angle of incidence}}{\text{sine of angle of refraction}}$$

There are a number of techniques available to determine refractive index based on this relationship.

The magnitude of the refractive index decreases with increasing wavelength, resulting in blue light being bent more than red when passing through a glass prism.

Spectral emission lines of various elements in the gaseous state are used to measure refractive index from the near-ultraviolet to the near-infrared portion of the spectrum. Table III lists the refractive indices for 13 wavelengths for three Schott glasses.[11] Approximate oxide compositions for these glasses (as analyzed at Corning Glass Works in 1969) are listed in Table IV. Figure 4 shows the change in refractive index with wavelength. As mentioned by Jenkins and White,[12] the important facts about the curves in Fig. 4 to be remembered are as follows:

(1) The refractive index increases as the wavelength decreases.
(2) The rate of increase becomes greater at shorter wavelengths.
(3) For different flint glasses, the curve at a given wavelength is steeper with a higher index of refraction.
(4) The curve for one flint glass cannot, in general, be obtained from that of another by a mere change in the scale of the ordinates.

The curves in Fig. 4 are known as dispersion curves. Dispersion is the rate of change of refractive index with wavelength ($dn/d\lambda$). It increases with decreasing wavelength and, for the flint glasses, increases with increasing refractive index.

The dispersion in optical catalogs is often defined as the difference in index between the F line and the C line of the hydrogen spectrum. These lines are

19

Table III. Refractive Indices of Selected Flint Glasses

Spectral line	Color	Wavelength (nm)	Element	Refractive indices LLF1	LF5	F2
i	Ultraviolet	365.01	Mercury	1.5793	1.6192	1.6662
h	Violet	404.66	Mercury	1.5691	1.6067	1.6506
g	Blue	435.84	Mercury	1.5633	1.5996	1.6420
F^1	Blue	479.99	Cadmium	1.5573	1.5923	1.6331
F	Blue	486.13	Hydrogen	1.5566	1.5915	1.6321
e	Green	546.07	Mercury	1.5510	1.5848	1.6241
d	Yellow	587.56	Helium	1.5481	1.5814	1.6200
D	Yellow	589.29*	Sodium	1.5480	1.5813	1.6199
C^1	Red	643.85	Cadmium	1.5451	1.5779	1.6158
C	Red	656.27	Hydrogen	1.5446	1.5772	1.6150
r	Red	706.52	Helium	1.5426	1.5749	1.6123
s	Infrared	852.11	Cesium	1.5384	1.5701	1.6067
t	Infrared	1013.98	Mercury	1.5354	1.5667	1.6028

*Center of the double line.

Table IV. Oxide Compositions (wt%) of Selected Flint Glasses

Oxides	LLF1	LF5	F2	SF2	SF4	SF6
SiO_2	60.8	52.3	45.6	41.5	30.8	27.0
Na_2O	4.6	7.0	3.5	1.0	1.0	0.5
K_2O	9.0	6.0	5.1	6.4	2.1	1.1
PbO	25.7	33.7	45.1	50.3	65.6	70.5
As_2O_3	0.4	0.5	0.4	0.3	0.2	0.5

rarely used these days due to the cumbersome nature of hydrogen sources and the convenience of the cadmium lamp.

For designing achromatic lenses, it is necessary to know the relationship between dispersion and refraction. The Abbé number (nu value) is a fraction commonly used in optical calculations that gives the reciprocal of a dispersion in relation to the refraction:

$$\nu_e = (n_e - 1)/(n_{F^1} - n_{C^1})$$

The relationship between the refractive index near the center of the visual spectrum (n_d) and the nu value (ν_d) is commonly used to characterize optical glasses.

For the flint glasses mentioned above the refractive index (n_D) and the nu value are listed in Table V. These values are plotted in Fig. 5. Note that ν_D increases from right to left. A plot of this sort is called an optical glass diagram.

Different manufacturers and different nationalities use different terminologies to describe optical glasses. For flint glasses the range of an individual

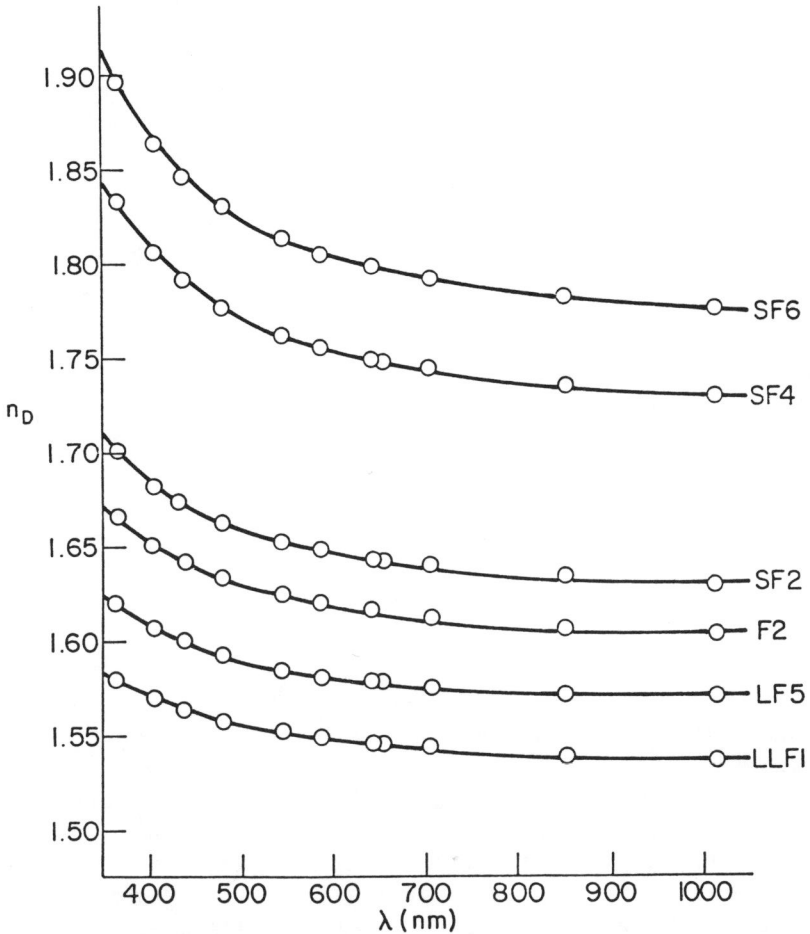

Fig. 4. Dispersion curves of selected flint glass.

Table V. Optical Properties of Flint Glass

Glass	n_D	ν_D
*LLF*1	1.548	45.8
*LF*5	1.581	40.9
*F*2	1.620	36.4
*SF*2	1.648	33.9
*SF*4	1.755	27.6
*SF*6	1.805	25.4

terminology is bounded by discrete ν values. In Fig. 5 the English (Chance) terminology is shown at the bottom, the German (Schott) at the top.

The optical properties (n_D, ν) of all alkali–lead–silicates will be very close to the line or extension thereof for the six flint glasses in Fig. 4. To illustrate,

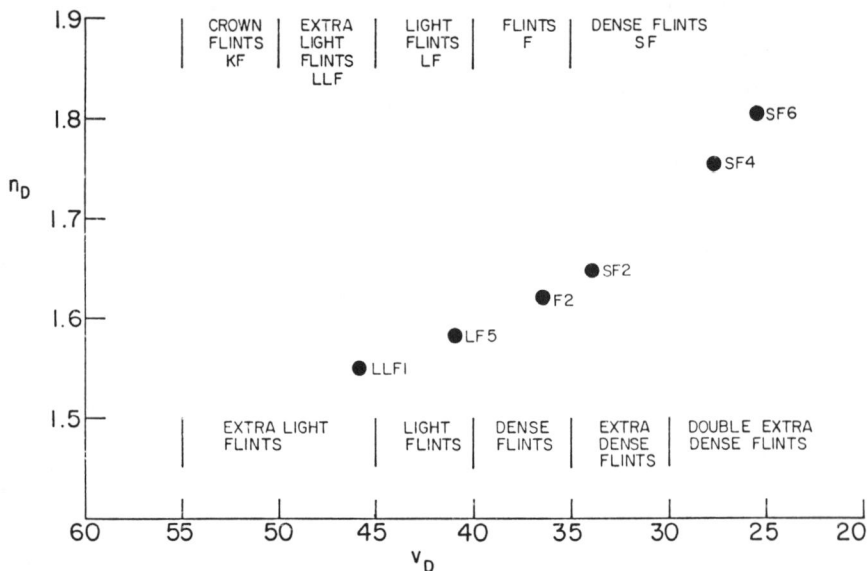

Fig. 5. Optical glass diagram for flint glasses.

the data of Merwin and Anderson (Table VI), as reported by Morey,[13] are shown in Fig. 6.

The positions of the flint glasses mentioned earlier are indicated by the filled circles. The oxide compositions, optical properties, and densities are listed in Table VI. The effect of the percentage of lead oxide on refractive index is plotted in Figure 7, both on a weight percent and a mole percent basis. On a weight percent basis, the effect is exponential, being more pronounced at high weight percentages of lead oxide. However, on a mole percentage the effect is nearly linear and probably would be if the mole percentage of alkali oxide were constant.

The following is a quotation from the preface to the first trade catalog of the Jena Glass Laboratory, issued in July 1886:[14]

"As a consequence of the uniformity of their chemical constitution, the silicate glasses hitherto in use could be arranged as a single series, in which, from the lightest crown to the densest flint (with some trifling exceptions) the dispersion steadily increased with the index.

A theoretical discussion of optical problems places it beyond doubt that the construction of instruments to fulfill simultaneously several given conditions would be greatly facilitated, if the optician had at his disposal glasses of the same mean index and various dispersions, and glasses of the same dispersion with various indices. It is, therefore, an important step in advance — that the systematic use of a larger number of chemical elements in the composition of glass has rendered such graduations possible, and that, in several instances, the choice between available glasses, instead of being substantially of a linear character as heretofore, has become *two-dimensional*."

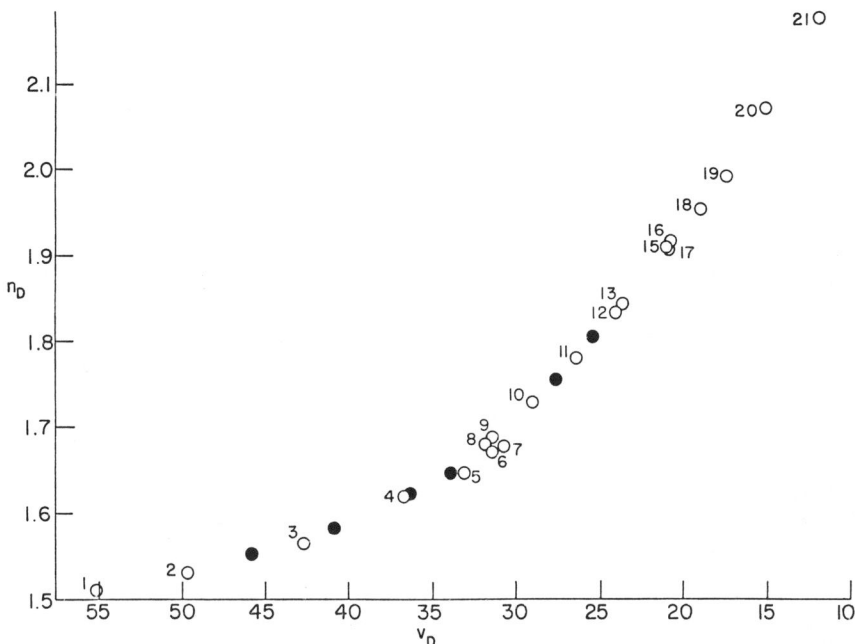

Fig. 6. Optical glass diagram for flint glasses according to Merwin and Anderson (in Ref. 13).

The 1886 list of optical glasses in the Jena catalog contained 44 glasses, of which about 18 did not lie on the standard crown–flint dispersion-index curve. Eight of these contained lead oxide!

Besides silica, the only other important glass-making acid oxides are phosphoric acid and boric acid, so the scientists at Jena tested the optical properties of glasses made from phosphoric acid and boric acid in combination with as many metallic oxides as possible. Lead oxide with boric acid, with or without the presence of silica, forms a glass readily. However, large regions of the $PbO-SiO_2-P_2O_5$ composition field have only two-phase opal glasses.

The lead borate glasses, though, were quite useful since boric acid is peculiar in shortening the blue partial dispersion relative to the red. These "short" flint glasses are still used in specialized lens systems to minimize the secondary spectrum. A typical composition from Schott and Genossen, Jena, is listed in Table VII. These lead borate glasses are particularly susceptible to chemical attack by moisture as well as by acidic or basic solutions such as are used in polishing or cleaning operations. Accordingly, they must be handled with great care and are only used for very specialized applications.

In spite of their low melting temperatures, it was difficult to obtain high homogeneity in the lead borate glasses because of the high volatility of the major constituents. Indeed, Michael Faraday struggled with this problem as early as 1830.[15]

23

Table VI. Optical Properties and Densities of Alkali Lead Silicate Glass

No.	Composition (wt%)				Composition (mol%)				n_D	ν_D	Density (g/cm³)
	SiO₂	Na₂O	K₂O	PbO	SiO₂	Na₂O	K₂O	PbO			
1	70.5	0.3	20.0	9.2	82.0	0.3	14.8	2.9	1.5071	55.1	
2	65.2	0.3	14.7	19.4	81.4	0.4	11.7	6.5	1.5286	49.7	
3	55.1	0.4	15.1	29.0	75.6	0.5	13.2	10.7	1.5629	42.7	
4	45.0	0.8	10.2	43.8	70.2	1.2	10.2	18.4	1.6157	36.4	
5	39.7	1.2	10.0	49.3	65.6	1.9	10.5	21.9	1.6450	33.2	3.553
6	33.1	9.4		52.5	62.1	14.8		23.0	1.6694	31.3	3.795
7	35.1	1.2	9.7	53.4	61.8	2.0	10.9	25.3	1.6754	30.7	4.023
8	40.4	1.5		58.1	70.3	2.5		27.2	1.6788	31.9	4.181
9	35.4	1.5	5.0	58.1	63.6	2.6	5.7	28.1	1.6836	31.4	4.190
10	34.4	1.8		63.8	64.5	3.3		32.2	1.7278	29.0	4.593
11	29.3	1.8		68.9	59.1	3.5		37.4	1.7800	26.4	5.001
12	24.9	1.6		72.9	53.9	3.4		42.5	1.8342	24.0	5.395
13	24.4	2.0	0.2	73.6	52.9	4.2	0.3	42.9	1.8415	23.7	5.450
15	20.0	1.6		77.9	47.0	3.6		49.3	1.9060	21.0	5.884
16	19.7	1.7		78.6	46.4	3.9		49.8	1.9152	20.7	5.944
17	20.5	0.4		78.9	47.4	3.8		48.8	1.9163	21.0	6.017
18	17.4	2.0		80.6	42.4	4.7		52.9	1.9543	18.9	6.170
19	15.2	2.3		82.6	38.3	5.6		56.1	1.9929	17.3	6.367
20	11.7	1.70		86.6	31.9	4.5		63.6	2.0789	15.0	6.84
21	8.0	1.83		90.04	23.5	5.2		71.3	2.179	11.9	7.21

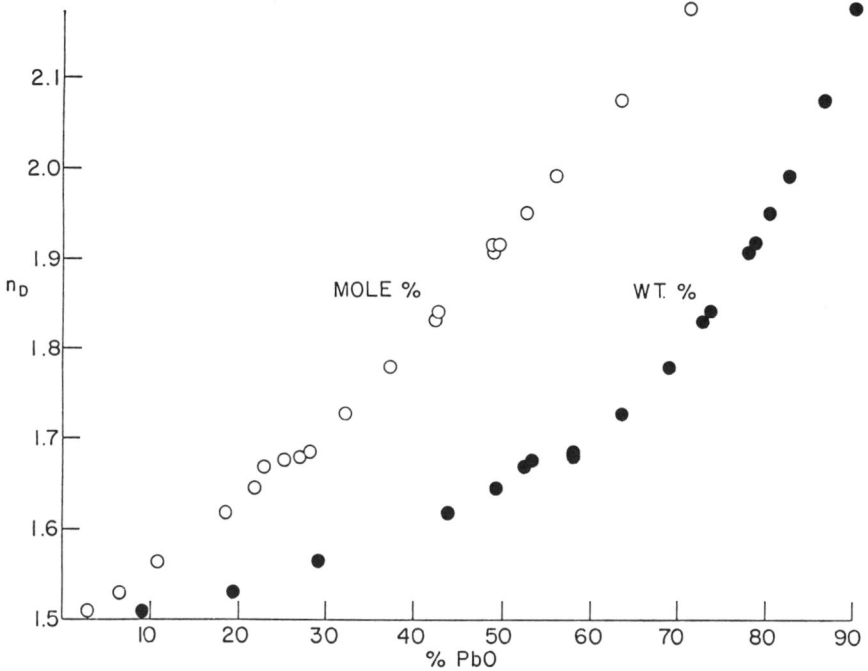

Fig. 7. Refractive index (n_D) as a function of lead oxide content for silicate glasses.

Table VII. Typical Composition* from Schott and Genossen

Oxides	Wt%
B_2O_3	52.7
Al_2O_3	9.0
PbO	38.0
As_2O_3	0.3

*n_D, 1.6131; ν_D, 44.0.

A supplement to the Jena catalog was issued in 1888 that contained eight new barium flint compositions that were especially designed for photographic lenses. These barium flints fulfilled the need for glasses which had the same index but different dispersions, or had the same dispersion but different indices in compositions which were relatively resistant to atmospheric or chemical attack. Essentially this was accomplished by substituting barium oxide and zinc oxide for lead oxide in the glass formulation. Table VIII gives some examples of glasses with approximately the same refractive index with different dispersions, while Table IX gives a pair of examples of glasses with the same dispersion but different indices. In recent years the barium flint glasses have played an important role in the ophthalmic industry, which is discussed in the next section.

25

Table VIII. Oxide Composition (%) of Barium-Flint Glasses of Similar Index

	Glass				
	LF5*	BaF3[†]	BaF6[†]	BaLF[†]	BaLF4[†]
n_D	1.581	1.583	1.589	1.589	1.580
ν_D	40.9	46.5	48.6	51.1	53.7
Oxides (wt%)					
SiO_2	52.3	50.2	48.6	46.7	48.7
B_2O_3			1.0	1.0	4.5
Al_2O_3				1.0	
Na_2O	4.6	1.0	0.5	1.9	2.5
K_2O	9.0	8.4	0.3	5.6	4.7
ZnO		7.6	8.4	11.0	15.7
BaO		13.6	20.5	24.7	20.8
PbO	25.7	18.7	14.2	7.6	1.0
TiO_2					1.1
As_2O_3	0.4	0.5	0.5	0.5	1.0

*Corning Glass Works Analysis, 1969.
[†]C. H. Hahner, "Optical Glass Manufacturing of Schott and Genossen of Jena," Combined Intelligence Objectives Sub-Committee Report, Item No. 9, File No. XXX11–22 (1945).

Table IX. Oxide Compositions* of Barium-Flint Glasses of Similar Dispersion

	Glass		
	BaF1	BaF6	BaF11
n_D	1.557	1.589	1.667
ν_D	48.6	48.6	48.4
Oxides (wt%)			
SiO_2	55.9	48.6	30.4
B_2O_3		1.0	9.3
Al_2O_3			0.6
Na_2O	2.0	0.5	
K_2O	13.0	6.3	
CaO			3.8
ZnO		8.4	5.2
BaO	11.5	20.5	43.6
PbO	17.3	14.2	2.9
Sb_2O_3			0.4
As_2O_3	0.3	0.5	0.4
TiO_2			3.4

*Compositions listed are from W. N. Wheat, "Schott & Genossen of Jena," Combined Intelligence Sub-Committee Report, Item No. 9, File No. XXXIII–69 (1945).

Ophthalmic Glasses

Ophthalmic glasses might be defined as optical glasses that are specially tailored to correct vision deficiencies. About 50% of ophthalmic glass is used for single-vision spectacles. For this need a standard lime glass crown with a refractive index of 1.523 is used. The remainder of the vision-correcting spectacles sold are bifocal or trifocal lenses.

Prior to 1906, three types of bifocal lenses were used:

(1) A separate lens held in position by a special frame attachment so that the lens could be moved to one side when not in use.
(2) A single lens with a different curvature on the reading portion, known as a one-piece bifocal.
(3) A segment of glass of different refractive index cemented onto the major crown glass.

In a fused bifocal lens the desired correction may be obtained by varying the contact curve between the crown glass and the high-index segment glass, as well as by varying the optical properties of the segment glass.

One of the earliest fused bifocal processes was that of the Kryptok Co.[16] In essence, it consisted of grinding a depression in the major crown portion having a desired curvature, placing a segment of higher index (either curved or flat) below and placing a wedge or ring between the two pieces of glass to let the entrapped air escape, heating the combination to an elevated temperature, which would allow the segment glass to sag and fuse to the crown glass in the process conforming to the curvature of the original depression. The wedge or ring would be subsequently ground off during the final grinding and polishing operation. The flint glass segments used were typically round and this type of bifocal was known for years as "Kryptoks."

Originally, three flint glasses were used, having indices of refraction of 1.616, 1.66, and 1.69 for the D line.[17] More recently (1960) a series of flint glasses having 10 different refractive indices were offered to the ophthalmic trade by Corning Glass Works. The objectives of this series of glasses were (1) to eliminate the need for thick segments for high ophthalmic corrections, and (2) to have a series of glasses all of which would be capable of satisfactory fusion to the standard white crown on the same thermal cycle.

From a glass composition standpoint, this series of glasses is a good example of how lead glass compositions can be manipulated to not only produce the required refractive index, but also control viscosity and thermal expansion. The approximate compositions and properties are listed in Table X. Apart from the refractive index, the lead glass must seal to the crown glass with a minimum of strain and the sealing (or fusing) temperature should be such that all glasses will fuse properly on the same schedule.

For this type of product it was determined that a good seal would occur when the flint glass reached a viscosity of about 10^6 P (10^5 N·s/m^2). Taking into consideration that the surface of the crown glass should be "tacky" yet not fluid enough to deteriorate the curvature of the countersunk deposition, the fusing

Table X. Compatible Flint Ophthalmic Glasses

	Corning code									
	8030	8031	8032	8033	8034	8035	8036	8037	8038	8039
n_D	1.567	1.582	1.597	1.611	1.626	1.641	1.656	1.670	1.685	1.700
ν_D	42.3	39.3	38.4	37.2	36.2	35.1	33.9	32.4	31.2	29.8
Oxide composition (wt%)										
SiO_2	55.8	52.1	48.8	45.3	43.6	41.9	39.8	38.0	36.4	34.7
Al_2O_3	2.0	2.0	2.0	2.0	2.0	2.0	2.0	2.0	2.0	2.0
Na_2O	11.0	7.9	5.5	3.4	4.1	4.4	4.7	2.8	0.9	
K_2O		2.1	3.7	5.4	4.6	4.1	3.4	5.3	6.9	7.6
PbO	31.0	35.7	39.8	43.7	44.0	44.6	45.9	47.4	49.0	50.7
TiO_2					1.5	0.5	1.0	2.0	3.0	3.8
ZrO_2						2.3	3.0	2.3	1.6	1.0
As_2O_3	0.1	0.1	0.1	0.1	0.1	0.1	0.1	0.1	0.1	0.1
Sb_2O_3	0.1	0.1	0.1	0.1	0.1	0.1	0.1	0.1	0.1	0.1
Physical properties										
Expansion ($\times 10^{-7}$/°C)	89	89	89	88	88	88	87	87	87	87
Density (g/cm³)	3.12	3.26	3.40	3.53	3.63	3.70	3.79	3.86	3.93	4.02
Temperature at various viscosity levels (°C)										
10^2	1393			1380	1337	1300	1284	1255	1260	1215
10^3	1112	1112	1104	1105	1088	1064	1055	1034	1035	1000
10^4	933	930	923	928	923	908	905	890	888	867
10^5	809	806	801	809	808	798	801	787	787	771
10^6	719	718	714	721	725	717	723	712	714	705
10^7	654	645	647	655	658	654	661	651	658	659
Softening pt	614	610	611	612	613	613	615	618	621	617
Annealing pt	445	438	438	438	441	447	454	459	469	475
Strain pt	408	401	400	398	404	409	416	423	433	440

temperature was found to be about 710–720°C. (The crown glass has a softening point of about 725°C.)

In a series of flint glasses, as the refractive index increases, the lead oxide content of the glass also increases. This results in softer glasses if no attempt is made to produce a constant temperature for a viscosity of 10^6 P by adjusting the potash–soda ratio. In addition, as the lead oxide content increases, the viscosity vs temperature curve tends to steepen; i.e., the glass becomes "shorter."

Therefore, in order to compensate for the increased lead content, it is necessary to increase the potash–soda ratio in order to maintain the same temperature at 10^6 P. However, since the glasses get shorter with higher lead contents, the softening points of the glasses must increase with increasing lead content. As a consequence, the annealing and strain points also increase, necessitating a decrease in the 0–300°C thermal expansion coefficient to maintain a satisfactory seal stress to the crown glass.

Since acid durability (resistance to staining by grapefruit juice) is a necessary requirement for ophthalmic lenses and since acid durability, to a certain extent, is a function of the silica content of the glass, it is necessary to introduce other oxides (than lead oxide) which have a large effect on refractive index. These are titania and zirconia.

The use of titania is preferred over zirconia since it does not adversely affect the acid durability as much as zirconia. It also goes into solution easier and does not (in small amounts) affect the liquidus temperature. It has the disadvantage that it couples with any ferrous iron present to produce a yellow tinge in the glass. Titania also produces a larger increase in dispersion per index increment than does zirconia.

Zirconia has the disadvantage that it is difficult to put into solution. Above 3% by weight in a 1.70 index flint, zirconia may give rise to a cloudy opal glass. Special batch materials, handling, and melting techniques are necessary to keep the zirconia stone level in this type of glass within economical limits. Zirconia also increases the liquidus temperature when present in this type of glass to an extent greater than 2–3%.

For these reasons, zirconia is first introduced in the 1.626 member of the series, increases to the 1.656 member, and then decreases. Titania is first introduced in the 1.641 member with a steady increase throughout the rest of the series.

All fused bifocal lenses made from crown and flint glasses show a defect due to the difference in dispersions of the refractive indices of the two types of glass. This defect manifests itself as a color fringe. While it may be minimized by using thinner segments as in the compatible flint series above, it could be completely eliminated if the two glasses had the same dispersion.

This concept was patented as early as 1924[18] but was difficult to put into practice. Early attempts to produce low-dispersion, high-index, high-barium glasses were not very successful. Surface tarnishing was a primary problem as a product,[17] but these high-barium glasses also could not be melted to satisfactory quality, free from bubbles or stria from the pot refractories, due to their corrosive nature.

Table XI. Barium-Containing Ophthalmic Segment Glasses

	Corning code			
	8323	8080	8316	8078
n_D	1.588	1.617	1.653	1.700
ν_D	52.0	49.4	42.2	36.2
Oxide composition (wt%)				
SiO_2	50.3	43.0	39.0	34.0
B_2O_3	4.0	4.0	3.5	
Na_2O	11.5	8.5	8.3	6.0
PbO		2.2	13.8	24.0
BaO	16.0	26.0	20.1	20.5
CaO	5.0	3.3	5.0	1.0
ZnO	7.5	6.4		4.0
La_2O_3				2.5
TiO_2	2.5	2.6	2.9	3.0
ZrO_2	3.0	3.8	7.2	4.8
As_2O_3	0.1	0.1	0.1	0.1
Sb_2O_3	0.1	0.1	0.1	0.1
Physical properties				
Softening pt (°C)	686.0	688.0	686.0	677.0
Annealing pt (°C)	550.0	551.0	553.0	531.0
Strain pt (°C)	518.0	517.0	520.0	495.0
Expansion ($\times 10^{-7}$/°C)	92.0	92.0	91.0	92.0
Density (g/cm^3)	3.05	3.33	3.52	4.03

The development of the continuous melting process for optical glass during World War II paved the way for the commercial development of high-barium ophthalmic lenses.[19]

Even with improved melting facilities, satisfactory all-barium glasses could be produced only up to a refractive index of about 1.6. Both durability and devitrification problems necessitated the introduction of lead oxide into the glass formulation. The four Corning glasses listed in Table XI exemplify this trend. The oxide compositions are approximate and are only used to illustrate a trend. In the 1.70 index glass, it is necessary to add lanthanum oxide in order to have a completely satisfactory composition.[20]

While the dispersions of these glasses are less than those of the comparable flint glasses, they are still not as low as that of the crown glass (which has a nu value of 58.6). Some color fringes still persist where the vision correction requires high curvatures.

The barium flint glasses have substantially higher softening points and annealing points than do the flint glasses, thus requiring a higher thermal expansion coefficient for a satisfactory fusion to the crown glass.

A third use of lead oxide in ophthalmic ware is in the high-index bifocal segment composition used in combination with photochromic lenses.[21] One such

Table XII. High-Index Compositions for Sealing to Photochromic Glasses

	Example corresponding to Corning code			
	1 8082	2 8083	3 8087	4 8088
n_D	1.589	1.618	1.653	1.700
v_D	45.1	40.5	36.3	32.9
Oxide composition (wt%)				
SiO_2	46.4	40.6	35.6	31.3
PbO	27.2	35.8	44.4	44.4
B_2O_3	7.3	6.6	5.9	5.4
Al_2O_3	6.4	6.2	6.0	4.6
BaO	8.0	6.5	5.6	8.0
Na_2O	2.6	2.1	1.5	1.0
Li_2O	1.2	0.7	0.2	0.2
TiO_2	0.6	0.9	1.2	2.8
La_2O_3				1.6
ZrO_2				0.5
Sb_2O_3	0.1	0.2	0.1	0.1
As_2O_3	0.2	0.2	0.2	0.2
Physical properties				
Softening pt (°C)	638.0	632.0	628.0	631.0
Annealing pt (°C)	484.0	486.0	482.0	500.0
Straining pt (°C)	454.0	454.0	450.0	470.0
Expansion ($\times 10^{-7}/$°C)	62.0	61.0	62.0	66.0

photochromic glass is designated Corning code 8111, marketed under the trade name Photogray Extra. This crown glass, index corrected to 1.523, possesses a softening point of 662°C, an annealing point of 495°C, and a coefficient of thermal expansion of $64 \times 10^{-7}/$°C. Because of the low expansion and low softening point of this glass compared to that of the standard ophthalmic crown, new ophthalmic segment compositions were developed. The key oxide to obtaining the high-index, low-expansion, low-softening-point compositions was that of lead. Compositions and physical properties of satisfactory compositions as disclosed in the patent are shown in Table XII.

Electrical and Electronic Glasses

The use of lead oxide containing glasses in the electrical and electronics industry dates back to 1879 when Corning Glass Works first made bulbs for Thomas A. Edison's incandescent lamp development. Corning had for years melted lead glasses for use in fine cut glass and other ware. A lead glass was particularly suitable for this application because of its long working range since the bulbs were formed by hand and subsequently needed to be sealed to a small evacuation tube at the top of the lamp bulb.

In addition, the thermal expansion closely matched that of platinum, which was used for the metal lead-in wires in the early lamps.

The use of lead glass in incandescent lamp bulbs continued for approximately 30 years. Originally the glass contained a mixture of sodium and potassium oxides; however, during World War I, when the supply of potash from Germany was cut off, a soda–lead glass was used. With the advent of automatic bulb-making machines in the 1920's, the composition of incandescent lamp envelopes was changed to a cheaper lime glass formulation. However, lead glass is still used for the stem, exhaust tubing, and flare because it has a higher electrical resistivity, is more easily worked, and is thermally softer than lime glass.

The fact that the electrical resistivity of the lead glass is higher than that of the lime glass is not as dependent on the presence of lead oxide in the glass as it is upon the concentration and nature of the alkali oxides present in the glass.

Compare the electrical properties of the glasses in Table XIII. Glass Code 0010 is probably very similar in composition to that used in Edison's lamp. Code 0050 is very similar to the bulb glass made during World War I. Code 0120 is the lead glass composition most widely used in the electrical and electronics industry. The oxide compositions listed are only approximate with minor, inconsequential ingredients such as fining agents omitted. The log of the resistivity (ρ) at 250 and 350°C is given for comparison purposes. Resistivity data are usually obtained in this range and are measured in ohm-inches. Note that the resistivity values for Code 0050 are very similar to the values of the lime glass composition used for lamp envelopes. The substitution of potash for soda in the formulation raises the resistivity, due to what is known as the "mixed alkali" effect.

While the composition of the glass for incandescent bulbs was changed to a lime glass base, lead glasses continued to be used for the envelopes of some radio tubes and power tubes in which a high electrical resistivity was needed. Compositions of the code 0120 type were most often used while the lower resistivity code 0010 type was used for neon sign tubing. The easy workability of a

Table XIII. Electrical Resistivities of Some Commercial Lead Glasses

	Glass Code		
	0010	0050	0120
Oxides (wt%)			
SiO_2	61.5	62	57
Al_2O_3	2	1	1.5
Na_2O	7	12.5	3.5
K_2O	7.5	0.5	9
PbO	22	24	29
Resistivity			
$Log_{\rho350}$	7.0	5.5	8.6
$Log_{\rho250}$	8.9	6.9	10.8

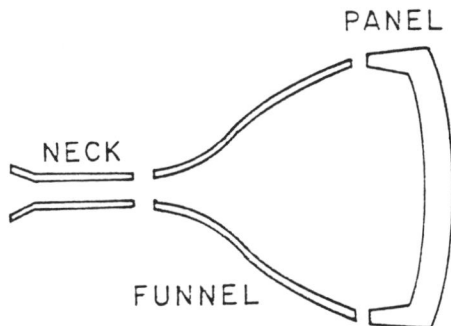

Fig. 8. Picture tube bulb components.

lead glass like code 0010 makes it ideal for forming the many intricate shapes in the neon tube industry.

The advent of radar during World War II saw the use of code 0120 type glasses expanded to make cathode ray tubes for radar applications. The early small cathode ray tubes were made by blowing in one piece; however, it was difficult to get the glass distribution and surface quality needed for long tubes by this method. A technique was developed in which the face plate or panel, the funnel, and the neck were made separately and these joined together (see Fig. 8).

The earliest television picture tubes were made of code 0120 type glass. However, the cost of lead oxide, the uncertainty of supply at that time, and the high density of the glass led to the development of a lead-free composition for the panel and funnel, although code 0120 type glass continued to be used for the neck.

With the advent of color television, higher voltages were used on the picture tube which increased the intensity of the X-rays generated inside the picture tube. New glass compositions were required. It was necessary to add about 9 wt% lead oxide to obtain the X-ray absorption needed. Lead oxide was not used in the panel because the greater thickness of the panel was sufficient for X-ray absorption, but more important, the presence of easily reducible oxides could be minimized in panel compositions to prevent browning of the face plate from high-energy electrons.

With even higher voltages being used in recent years, it has been necessary to increase the lead oxide content of the funnel composition to about 23% by weight of lead oxide as well as to add small amounts ($\sim$2 wt%) to the panel glass. With newer techniques of depositing phosphors and masking the area around the phosphors, a small amount of reducible oxides can be tolerated in the panel composition.

Another lead-containing glass, which is sometimes used in the electronics industry, is Corning code 7720, often referred to by the trade name Nonex. It is a low-expansion glass which will seal directly to tungsten and is a by-product of the development of easily melted, low-expansion borosilicate glasses used in railway signals. The approximate composition and pertinent physical properties are listed in Table XIV.

33

Table XIV. Composition and Properties of Code 7720 Glass

Oxides (wt%)		Physical properties	
SiO_2	74	Softening pt (°C)	755
Al_2O_3	1	Annealing pt (°C)	523
B_2O_3	15	Strain pt (°C)	484
Na_2O	4	Expansion ($\times 10^{-7}$/°C)	36
PbO	6	Density (g/cm^3)	2.35
		$\text{Log}_{\rho 350}$	7.3
		$\text{Log}_{\rho 250}$	8.8

References

[1]H. P. Rooksby, "Opacifiers in Opal Glasses Through the Ages," *GEC J. Sci. Technol.,* **29** [1] 20–26 (1962).

[2]W. E. S. Turner, "Studies in Ancient Glasses and Glass Making Processes: Part IV, The Chemical Composition of Ancient Glasses," *J. Soc. Glass Technol.,* **40**, 162–86 (1956).

[3]M. A. Besborodov, "A Chemical and Technological Study of Ancient Russian Glasses and Refractories," *J. Soc. Glass Technol.,* **41**, 168–84 (1957).

[4]F. J. Gooding and E. Meigh, Eds.; pp. 42–43 in Glass and W. E. S. Turner. The Society of Glass Technology, Sheffield, England, 1951.

[5]B. F. Biser; p. 100 in Elements of Glass and Glass Making. Glass and Pottery Publishing Co., Pittsburgh, PA, 1899.

[6]M. Schofield, "A Fruitful Period in Optical Glass," *Glass,* **48**, [12] 336–37 (1971).

[7]Anon., "Some Account of the Late M. Guinand and of the Important Discovery Made by Him in the Manufacture of Flint Glass," Longman, Hurst, Rees, Orme, Brown, and Green, London (1825).

[8]H. Volkmann, "Ernst Abbe and His Work," *Appl. Opt.,* **5** [11] 1720–31 (1966).

[9]F. W. Glaze and C. H. Hahner, "Optical Glass at the National Bureau of Standards," *Glass Ind.,* 562–67, 586–88, 590–91 (1948).

[10]R. W. Douglas and S. Frank; p. 80 in A History of Glass Making. G. T. Foulis & Co., Ltd., Henley-on-Thames, Oxfordshire, England, 1972.

[11]Schott Optical Glass Catalog, Jenaer Glaswerk Schott & Gen. Mainz, 1977.

[12]F. A. Jenkins and H. E. White; p. 466 in Fundamentals of Optics. McGraw-Hill, New York, 1957.

[13]G. W. Morey; p. 375 in The Properties of Glass, 2nd ed. Reinhold, New York, 1954.

[14]H. Hovestadt; p. 6 in Jena Glass and Its Scientific and Industrial Applications. Translated and Edited by J. D. Everett and A. Everett. MacMillan, New York, 1902.

[15]M. Faraday, "On the Manufacture of Glass for Optical Purposes," *Philos Trans,* 1–37 (1830).

[16]C. F. Deickman, "Method of Making Bifocal Lenses," U.S. Patent No. 865 363, 1907.

[17]R. J. Montgomery, "The Manufacture of Fused Bifocal Spectacle Lenses," *J. Am. Ceram. Soc.,* **12** [4] 274–303 (1929).

[18]T. B. Drescher, "Bifocal Lens," U.S. Patent No. 1 489 630, 1924.

[19]W. H. Armistead, "Ophthalmic Glass," U.S. Patent Nos. 2 523 624; 2 523 625; 2 523 626; 1950.

[20]G. B. Hares and D. W. Morgan, "Barium Flint Ophthalmic Glasses," U.S. Patent No. 3 902 910, 1975.

[21]G. B. Hares and D. L. Morse, "High Index Ophthalmic Glasses," U.S. Patent No. 4 211 569, 1980.

Electrical Characteristics of Glasses

The DC resistivity of all glasses is an important electrical property that indicates trends in the elevated temperature loss tangents, the thermal dielectric strength values, and the values of the low-frequency loss tangent at room temperature.

The DC resistivity of glasses is a function of the alkali content — the ionic carriers — and the degree of opposition the glass structure offers to the migration of these ions through the glass. If the structure is "tight," then the opposition is large and the DC resistance will be high. If the concentration of the alkali ions is low and the opposition is high, then the DC resistance will be even higher. The opposition is high if the number of modifying ions is large.

The loss tangent of glasses at room temperature and between 10^2 and 10^4 Hz is a function of the number of alkali ions and the glass structure. If the alkali content is low and the structure is "tight," then the loss tangent will be low. The loss tangent at microwave frequencies is primarily a function of the heavy modifier ions. The dielectric constant increases with an increase in glass density. If the heavy ion content is large, then the dielectric constant will be large. However, the dielectric constant decreases at microwave frequencies due to the increase in absorption.

Electrical Properties of Lead Glasses

The electrical properties of lead glasses are, in general, superior to those of other glasses. The DC resistivity can be very high, especially near room temperature, and is more than adequate for most applications. The loss tangent values below the microwave frequencies are low, and some of the glasses show low losses even as high as 1 GHz.

Lead in glasses, except for very high concentrations, is a network modifier and as such tends to "block" the motion of the alkali ions. This increases the DC resistivity and reduces the dielectric loss tangent. The lead ions increase the glass density and the dielectric constant. The lead also increases the dielectric losses at microwave frequencies.

At very high concentrations, lead ions begin to act as network formers. This changes the electrical trends established at lower concentrations and prevents the attainment of very low losses and high DC resistivity values. This transition starts near the composition $PbO \cdot SiO_2$.

The compositions of several Corning Glass Works commercial glasses discussed in this section are shown in Table I. The lead content varies from 20 to 80% and represents a reasonable cross section of lead glasses.

Resistivity

The DC resistivity of glasses follows the relationship:

$$\log \rho = A + B(1000/K)$$

where ρ = DC resistivity, K = temperature in degrees Kelvin, and A and B are constants.

Table I. Composition of Several Commercial Glasses

Glass code	Density	K^1 (1 kHz)	SiO_2	Al_2O_3	B_2O_3	PbO	K_2O	Na_2O	Li_2O
0010	2.79	6.7	63	1		22	6	8	
0120	3.05	6.75	56	2		29	9	4	
1990	3.5	8.3	41			40	12	5	2
8871	3.84	8.45	40.5			49.8	5.6	2.7	1.0
8161	3.99	8.3	39			51	6		
0090	4.18	9.15	38.6			57.9	1.57		
1070	5.42	15	3	11	12	74			
8363	6.27	17	5	3	10	82			

The value for B is related to the glass alkali content and space available to it for migrating through the glass. The alkali content is known from the composition, and the migration space availability can be derived from the composition by determining the glass structure and knowing the type and number of the network forming and modifying ions. A formula[1] developed by J. M. Stevels is one way of making these calculations.

The value of A is primarily influenced by the glass alkali content and glass density. The A value also can be estimated from glass composition.

The curves in Figs. 1 and 2 show the logarithm of the DC resistivity plotted as a function of $1000/K$. The sequence of the curves in Fig. 1 is fairly close to that expected from the glass compositions. The two curves in Fig. 2 do not follow this sequence, and this shift from an increase in DC resistivity with an increase in lead content is shown in Fig. 3 for the glasses with 74 and 82% lead content. Instead of the anticipated increased resistivity, there is a reversal and the resistivity decreases. This is an indication that some of the lead ions are acting as network-forming ions, thereby opening the network and reducing the alkali "blocking." It can be estimated that about one-sixth of the lead for code 7570 is acting in this capacity and an even higher percentage assumes this role in Corning code 8363 glass. This transition occurs at different percentages depending upon the type and number of the other glass network formers as well as the presence of other network modifying ions. This dual role of the lead ion prevents lead glasses from obtaining DC resistivity characteristics that are extremely high. The DC resistivity properties are, however, as good as and even better than most commercial glasses.

Dielectric Constant

The dielectric constant at room temperature and 10^3 Hz increases with an increase in density. Since lead is a heavy ion, the glass density is high and so is the dielectric constant. The curve in Fig. 4 shows this relationship for the group of glasses listed in Table I. The dielectric constant can be plotted as a function of the percent of PbO, as shown in Fig. 5. The relationship seems to be linear for the series of lead–alkali–silicate glasses, but the lead–boroalumina–silicate glasses show a departure from this trend. It is expected

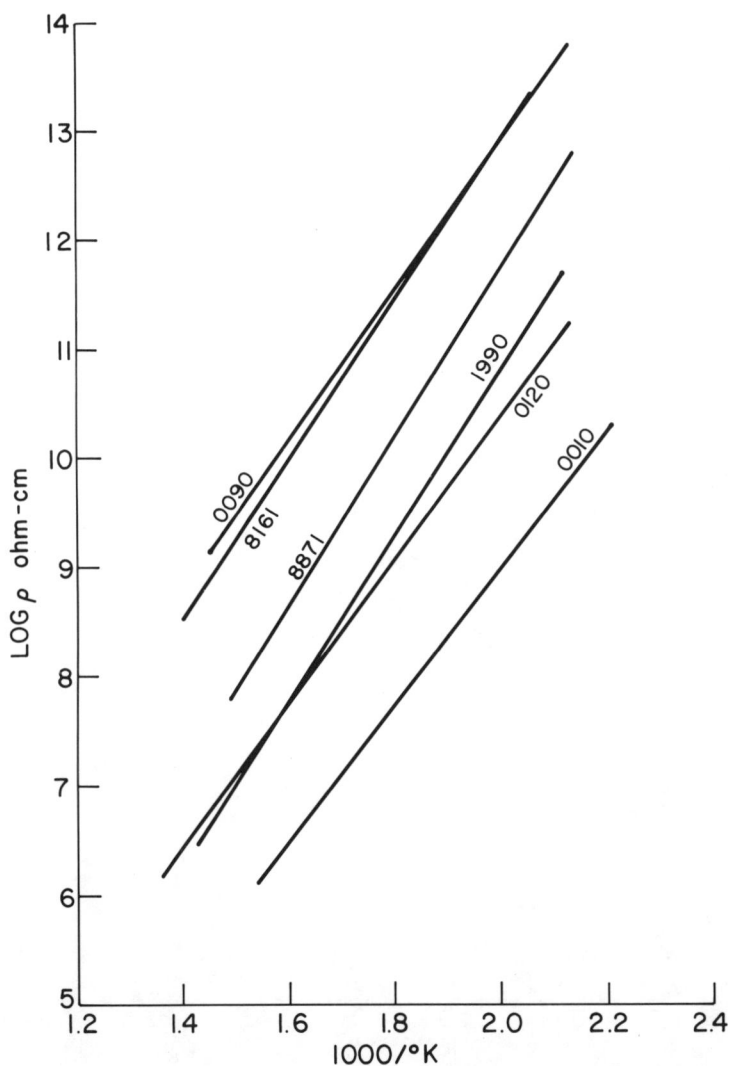

Fig. 1. Logarithm of the DC resistivity in ohm·cm as a function of the reciprocal of the absolute temperature for several Corning code glasses.

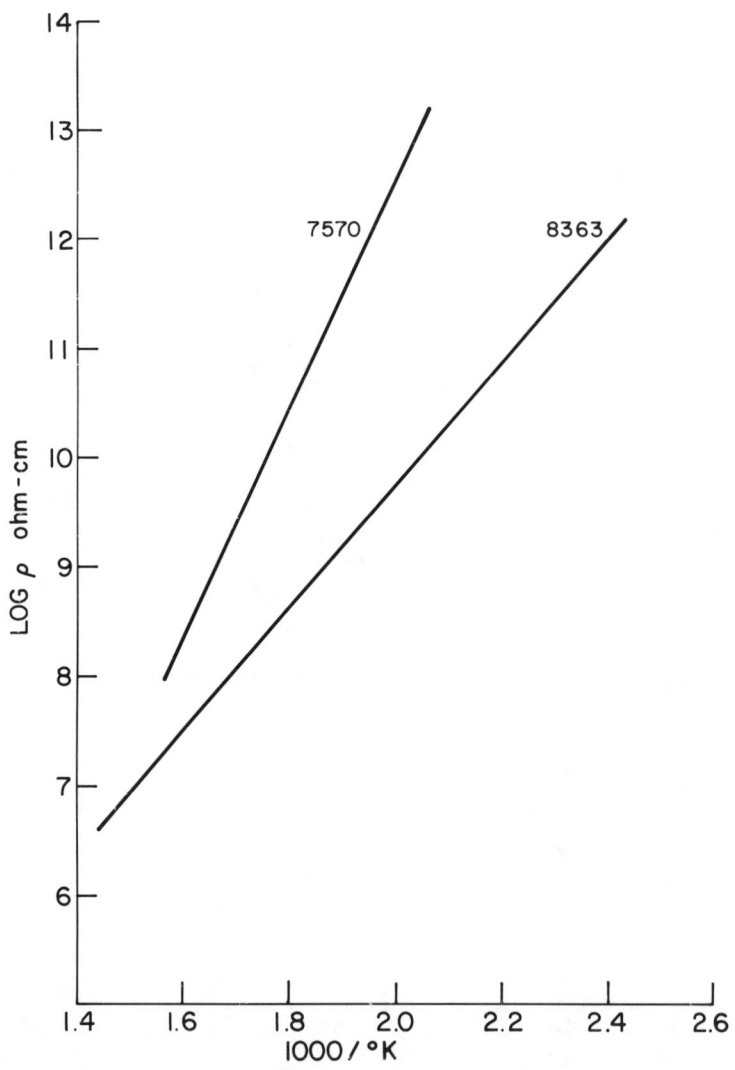

Fig. 2. Logarithm of the DC resistivity in ohm·cm as a function of the reciprocal of the absolute temperature for two Corning code glasses.

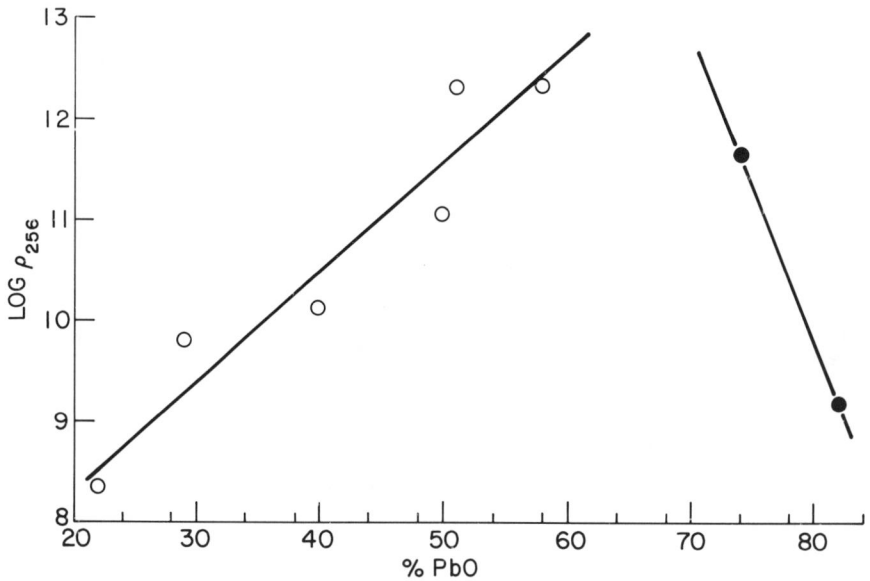

Fig. 3. Logarithm of the DC resistivity in ohm·cm at 250°C as a function of the lead weight percentage for several Corning code glasses.

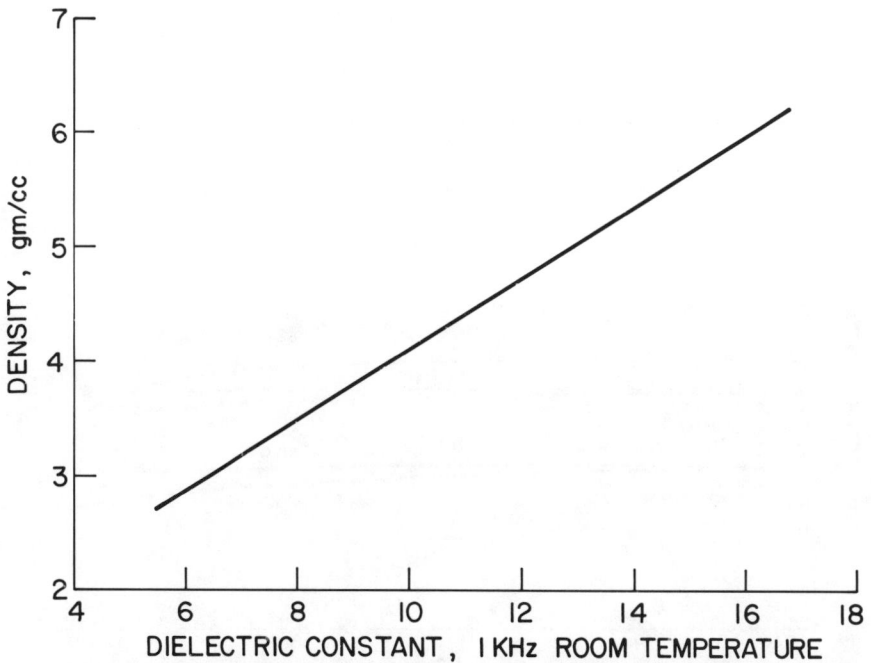

Fig. 4. Dielectric constant at 1 kHz and room temperature as a function of glass density for several Corning code glasses.

39

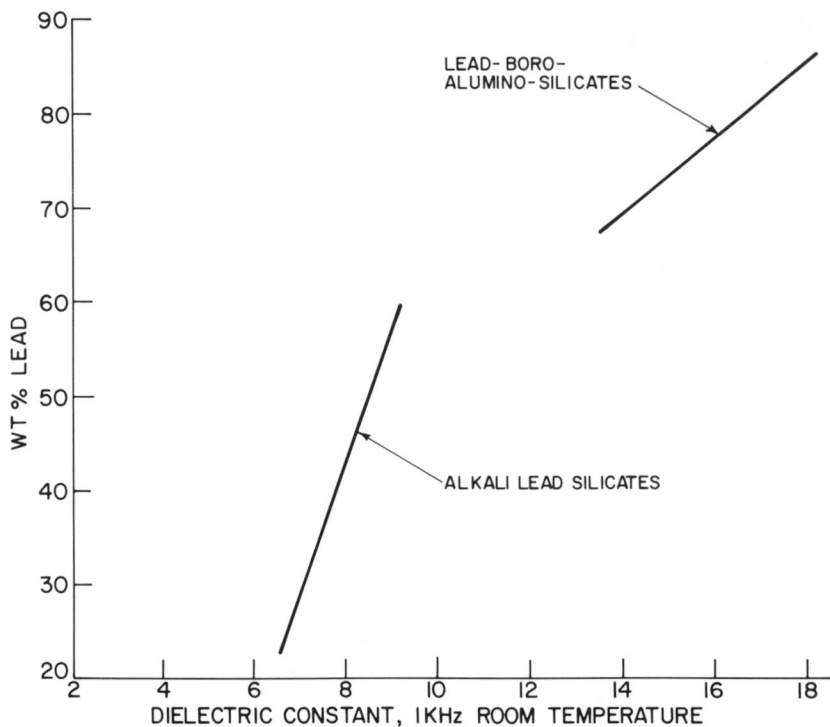

Fig. 5. Dielectric constant at 1 kHz and room temperature as a function of lead wt% for several Corning code glasses.

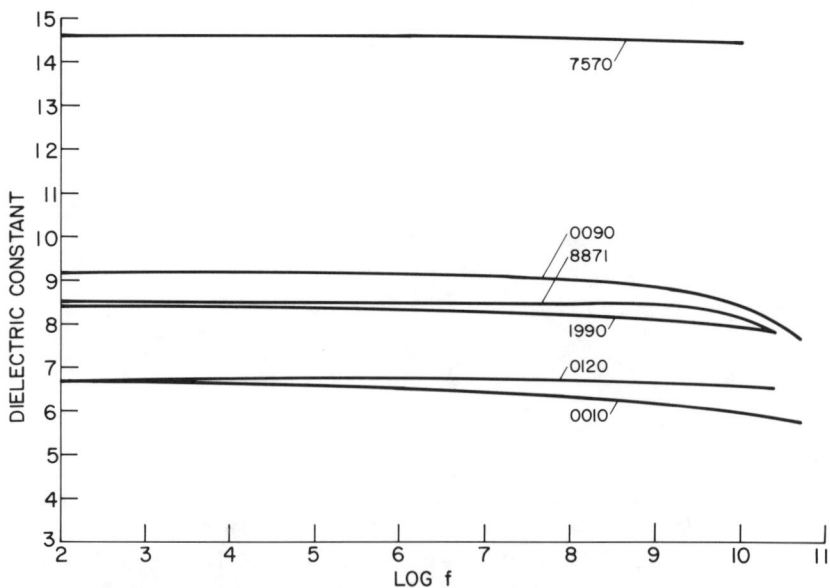

Fig. 6. Dielectric constant at room temperature as a function of frequency for several Corning code glasses.

40

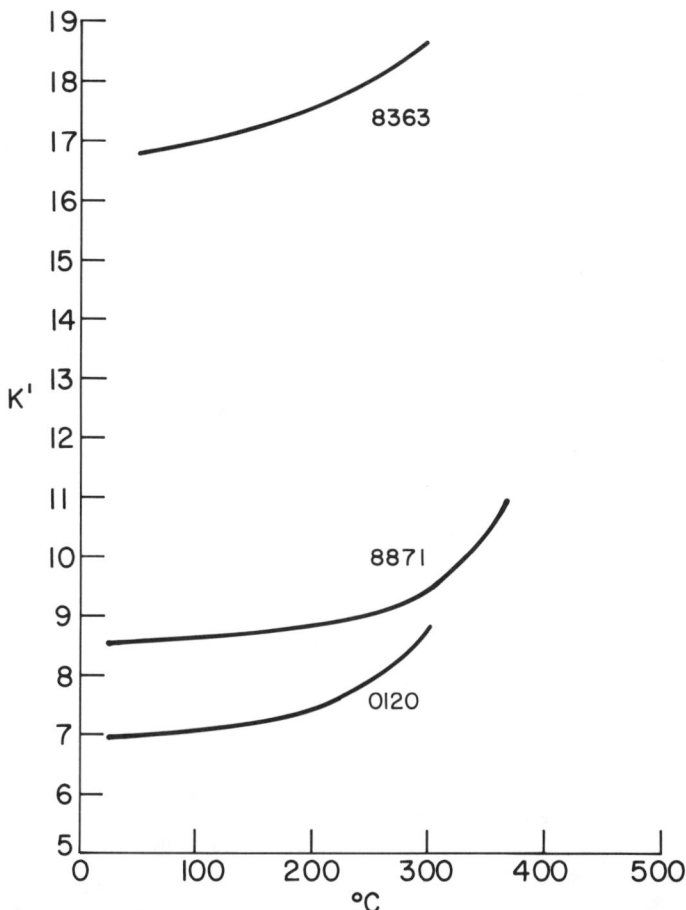

Fig. 7. Dielectric constant at 10^4 Hz as a function of temperature for three Corning code glasses.

that the B_2O_3 and Al_2O_3 also contribute to the dielectric constant, thereby indicating a different compositional response.

The curves in Fig. 6 show the response of dielectric constant as a function of frequency at room temperature. These curves show that the dielectric constant is relatively insensitive to frequency except in the microwave bands. Here the dielectric constant decreases with an increase in frequency and is an indication of an absorption that is primarily a function of the content of the heavy modifier ions.

The low-frequency response is influenced by the high-frequency portion of an absorption attributed to the restricted motion of alkali ions.

The curves in Fig. 7 show that the dielectric constant remains the same as a function of temperature up to approximately 250°C and then increases fairly rapidly. The relative values of DC resistivity indicate the temperature where this increase occurs.

41

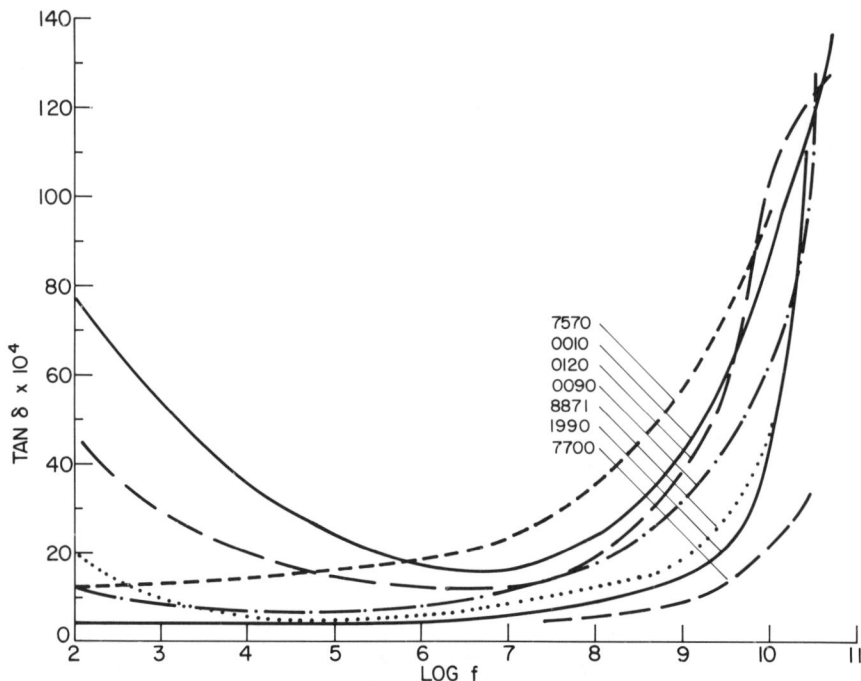

Fig. 8. Loss tangent at room temperature as a function of frequency for several Corning code glasses.

Dielectric Loss Tangent

The low-frequency loss tangents in Fig. 8 show that the losses vary inversely with the DC resistivity. Glasses with high DC resistivity values tend to show low-frequency loss tangent values which are low. The same type of structure that restricts the movement of alkali ions through the network also restricts the oscillatory motion of the same ions. This restriction reduces the loss tangent values at the lower frequencies. As is the case with the dielectric constant, discussed above, this is the high-frequency part of an absorption governed by alkali movements. At higher frequencies the heavy ion absorption predominates, and the loss tangent values rapidly increase. In Fig. 8 Corning code 7070 glass is also plotted, and the loss values of this glass are very low at microwave frequencies. This glass is an alkaline borosilicate, and because there are no heavy modifying ions in the composition, the loss values are low.

The curves in Fig. 9 show the loss response as a function of temperature. As with the dielectric constant, these values tend to be constant up to about 100°C. Above 200°C the losses of most lead glasses tend to increase, since lead glasses tend to soften at temperatures lower than those of other glasses and the restriction to alkali movement is reduced at relatively low temperatures.

Dielectric Strength

In most practical applications involving electrical insulation, there are two types of dielectric failures. Most failures that occur at room temperature are a

42

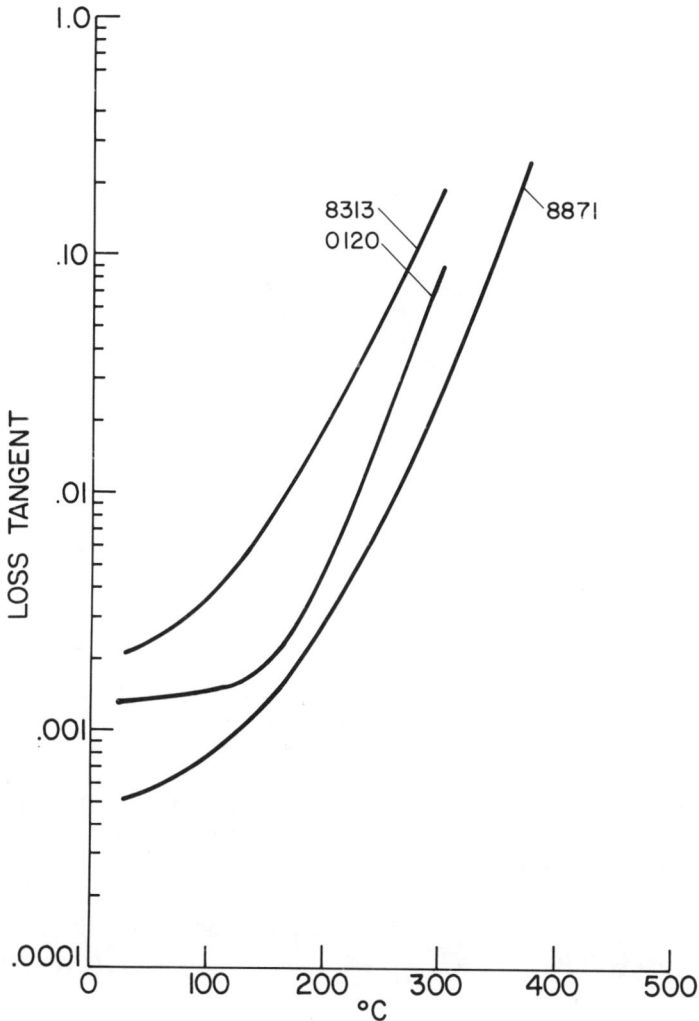

Fig. 9. Loss tangent at 10^4 Hz as a function of temperature for three Corning code glasses.

result of corona deterioration. The configuration of the electrical contact to the insulating material and the insulation thickness, dielectric constant, and atmospheric environment are all contributing factors for corona development and subsequent deterioration of the insulation.

The other type of dielectric failure is a thermal reaction where the glass absorbs energy, and both its temperature and loss tangent increase, leading in some cases to cascade failure.

The curves in Fig. 10 show both types of failures for a few of the glasses. At the lower temperatures, the corona failures are almost independent of the glass type. At increased temperatures the thermal failures predominate, and the

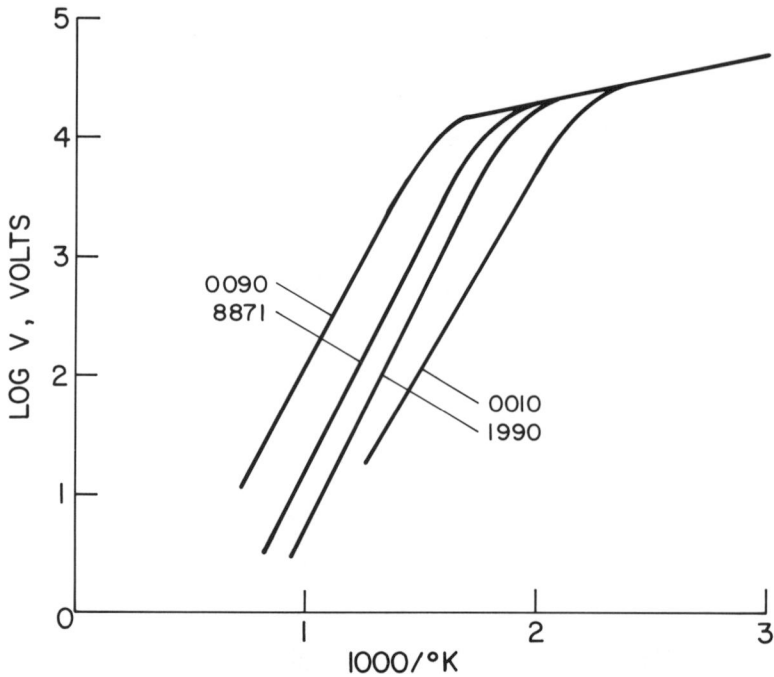

Fig. 10. Logarithm of the breakdown voltages of 2 mm thick samples as a function of the reciprocal of the absolute temperature for a few of Corning code glasses.

sequence of these failures follows closely the sequence of their DC resistivity curves. The DC resistivity is the predominant loss mechanism for these materials at the higher temperatures.

The curves in Fig. 11 show a similar response of Corning code 0120 glass for various sample thickness. Both the corona and thermal failures require higher voltages as the sample thickness is increased. A similar response can be shown for the other glasses of this section.

Summary

Lead glasses have high dielectric constants, low loss tangent values for frequencies near 10^6 Hz, and high loss tangent values at microwave frequencies. At high concentrations lead changes its role in the glass structure, limiting the attainment of extremely high DC resistivities.

The overall electrical characteristics of the better lead electrical glasses are exceptionally good.

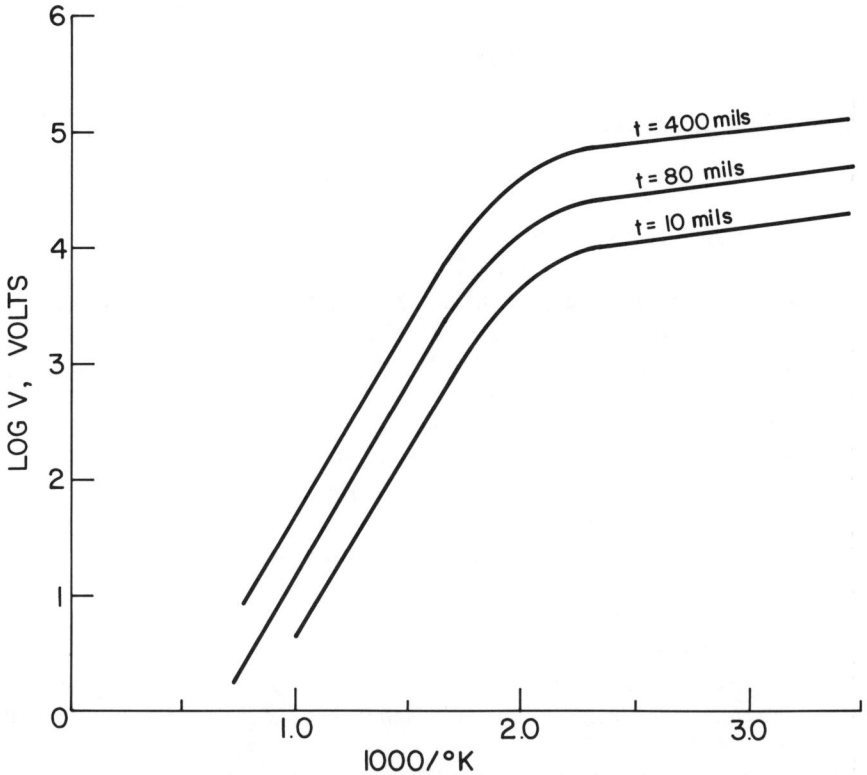

Fig. 11. Logarithm of the breakdown voltages as a function of the reciprocal of the absolute temperature for various thicknesses of Corning code 0120 glass.

References

[1] J. M. Stevels; Progress in the Theory of the Physical Properties of Glass. Elsevier, New York, 1948.

Solder Glasses

Stable Solder Glasses

The term solder glass is used to define a soft glass that can be used to join conventional glasses at temperatures low enough so that significant distortion of the glass parts is avoided. The actual sealing is usually done in a furnace.

In some ways the term is deceptive in that, unlike metal solders, the solder glass does not become extremely fluid and flow into joints as do metallic solders; neither are parts heated with a torch or a hot soldering iron.

Corning Code 7570 was developed by Dalton[1] for use with soda lime glass. It contains about 75% lead oxide and has a thermal expansion (0–300°C) of 82×10^{-7} cm·cm^{-1}/°C and a softening point of 440°C. The expansion to set point is some 15% higher so it matches a conventional glass having an (0–300°C) expansion of 90×10^{-7} cm·cm^{-1}/°C. Sealing is typically at 550°C for 15–30 min.

In the rest of this section, physical properties such as softening point and expansion of the solder glass will not be given, but rather the expansion range of materials that can be sealed and the required temperature and time for sealing. The expansion will be in the 10^{-7} cm·cm^{-1}/°C units customarily used in describing glasses. On this scale, lime glasses are 90, Kovar sealing glasses around 50, tungsten sealing glasses around 40, Chemical Pyrex code 7740 32, and fused quartz 6.

Attempts at sealing lower expansion, low-silicate glasses were less successful. They are often described as hard glasses, but this refers to their flame-working characteristics; their distortion temperature limits are not much higher than lime glasses save for the aluminosilicate types.

Code 1826 glass was designed for use with Kovar sealing glasses but was somewhat too high in expansion and required too high a sealing temperature (680°C) and thus never found much application. It has been used to seal single-crystal sapphire windows to appropriate glasses.[2]

The solder-glass seal is usually of the butt-joint variety, with the solder glass applied to one or both surfaces of the parts to be joined. The solder glass is most often applied as a powdered glass made into a thick slurry with a viscous binder (1.5% nitrocellulose dissolved in amyl acetate is often used). Other techniques involve dipping parts in a slurry or a spray application. Hot-glass application and the use of preforms have also been attempted.[3]

Because a good fit of the parts to be joined is desirable, parts are often mechanically ground. It is then necessary to acid etch in hydrogen fluoride or ammonium bifluoride solution to remove potential break sources.

One of the unique properties of solder glass is its ability to make a seal where a direct fusion seal is not possible as, for example, in glass-to-mica seals. Solder glasses bond well to aluminum. Seals to an aluminum sheet up to 8- mil $(2.54 \times 10^{-5}$ m) thickness are possible. In this case, the aluminum does not match in expansion but is sufficiently ductile.

Another example is a seal between a silver-plated sealing alloy and appropriate glass via a solder glass.

Devitrifying Solder Glasses

The most important application, by far, for solder glasses is in sealing together the parts of a color television bulb. The first color television bulbs were round and used a metal funnel with a glass face plate. The funnel was broken into two parts with flanged sections that were subsequently welded together after the aperture mask assembly had been inserted and mounted.

Much work was done on a bulb with glass panel and glass funnel both sealed to metal flanges that could be welded for closure. The metal, in this case, was a titanium-stabilized, 17% chrome iron, commonly known as 430 Ti. The matching glass had an expansion of 100.

At the same time, attempts were made to produce a solder glass that could seal the glass parts directly at temperatures not exceeding 450°C. Some of the more promising candidates showed a strong tendency to devitrify in the sealing cycle. Eventually, it was realized that this could be used to advantage.[4] As it turned out, a stable solder glass would have been defective on two counts, because the manufacturing process also included a vacuum bake at 400°C. The solder-glass parts exposed to the vacuum would froth by expansion of internal gas bubbles, and in addition, the solder glass would offer little resistance to the independent deformation of the glass parts under vacuum load. The independent deflection of the glass parts would be frozen in by the solder glass and the joint area under high stress and would almost certainly fracture if the bulb were subsequently returned to atmospheric pressure.

Code 7520 was the first devitrifying solder glass used in color-tube production. It required 440°C sealing temperature for 1 h. It worked well with round TV bulbs but with the advent of rectangular designs was replaced by other compositions that set up faster and developed greater rigidity in the sealing process.

An essential ingredient in these glasses is zinc oxide. Control of the crystallization rate is governed by the SiO_2 content, which is quite critical.

The crystallization is complex, but the initial crystallization seems to be that of a ternary $2PbO \cdot ZnO \cdot B_2O_3$ which has been described by Petzoldt.[5] This is followed by lead borate phases. The massicot form of lead oxide has been observed in the devitrified product in some glasses.

In addition to color TV bulbs, the devitrifying solder glasses have found other applications, such as in joining a glass faceplate to a ceramic funnel (Forsterite) for display tubes. It has been used to join alumina parts in integrated circuit packages.

Some modification downward in expansion is possible by adding low-expansion ceramic materials in the powdered glass. Zircon has been quite useful in this regard.[6]

Code 7578 is a relative of the TV solder glasses but has a much higher ZnO content. It seals at 520°C in 0.5 h and matches a glass of 65 expansion. Another ternary lead zinc borate ($PbO \cdot 2ZnO \cdot B_2O_3$), also described by Petzoldt,[5] is a principal phase. This glass does not set up as rigidly at the higher sealing temperature as do the TV sealing glasses. It was used at one time for joining a fiber-optic faceplate to a 44% nickel iron flange for use in infrared sniper scopes.

Code 7574 is a zinc borosilicate composition that crystallizes to a mixture of zinc borate and zinc silicate crystals. Since code 7574 is free of reducible oxides, it can be used in a neutral or slightly reducing atmosphere. It has been used to seal aluminosilicate glasses in the 40 expansion range at 750°C/h.[7] It also has application for tungsten and molybdenum sealing.

The solder glasses described above have very poor acid durability. They can, in fact, be completely dissolved in nitric acid. This is convenient for reclaim operations. A 3N nitric acid solution at 60°C is optimum for solder-glass removal.

Although readily attacked by acid solutions, the solder glasses have reasonable durability toward ordinary moisture attack and, being alkali free, maintain a good electrical surface resistivity.

Codes 7594 and 7595 are essentially lead borosilicates with a high content of TiO_2. In this case, a single phase is precipitated, namely, lead titanate. Lead titanate has a negative expansion so that the devitrified glasses have a low expansion even with a relatively high residual glass content. Although the material is soft enough to seal at lower temperatures, a temperature of 620°C for 0.5 h is required to assure proper phase development. At lower sealing temperatures, an undesirable metastable lead titanate phase is encountered.[8]

Code 7594 has been used to seal fritted disks of code 7740 to code 7440 tubing, a seal that would be difficult to do by direct flame sealing.

Because of its high glass content, it has been possible to make a seal between crystallized layers of code 7595 at temperatures lower than the original sealing temperature and in a very short time.[9]

Chemically durable lead titanate compositions have been developed for decorating low-expansion substrates, but these require a higher firing temperature (750°C).[10]

Glasses for Diode Encapsulation

Perhaps some mention should be made of diode-encapsulating glasses. In common with solder-glass applications, sealing is done in a furnace. Although a low sealing temperature is desired, the glass must be stable enough for continuous tube drawing. For minimum deterioration of the diodes to be encapsulated, the glass should be alkali free or at least soda free.

The silicon diode is assembled between two heavy metal leads (studs), and the cut tubing is slipped over and then shrunk around and sealed to the stud in a furnace operation.

Code 7063 is a lead borosilicate glass that has been used for diode sealing with molybdenum studs.

Code 8870 is a soft lead glass that has been used with dumet (copper-clad 42% nickel iron) studs. It is soda free but does contain potash.

48

References

[1]R. H. Dalton, "Solder Glass Sealing," *J. Am. Ceram. Soc.,* **39** [3] 109–112 (1956).

[2]F. W. Martin, "Sealing of Glass to Synthetic Sapphire," Proceedings of the Symposium on the Art of Glass Blowing, 1964.

[3]R. H. Dalton, "How to Design Glass-to-Metal Joints," *Prod. Eng. (N.Y.),* **36**, 26 (1965).

[4]Stewart A. Claypoole, "Composite Article and Method," U.S. Patent No. 2 889 952 (1959).

[5]Jurgen Petzoldt, "X-ray Analytical Studies in the Oxide System $PbO-ZnO-B_2O_3$," *Glastech. Ber.,* **39** [3] (1966).

[6]F. W. Martin and Frank Zimar, "Fusion Seals and their Production," U.S. Patent No. 3 258 350 (1966).

[7]F. W. Martin and Frank Zimar, "Properties and Applications of Devitrifying Solder Glass," Proceedings of the Sixth Symposium on the Art of Glass Blowing, 1961.

[8]F. W. Martin, "A Metastable Cubic Form of Lead Titanate Observed in Titania-Nucleated Glass-Ceramics," *Phys. Chem. Glasses,* **6**, [4] 143–6 (1965).

[9]F. W. Martin, "Processor Producing Seals," U.S. Patent No. 3 410 674 (1968).

[10]High Durability, Lead-Titanate-Containing Enamel for Glass-Ceramics, U.S. Patent No. 3 663 244 (1972).

Radiation Control Glasses

Glass and radiation science are historically connected. Roentgen's X rays of 1895 were produced by electron impingement on the glass envelope of his tube; thus, glass was the first source of X rays. The light emitted from the glass, accompanying the X rays, suggested to Becquerel a connection between the new X rays and the older fluorescence. His pursuit of this idea with fluorescent uranium components led to the discovery of radioactivity in 1896. Thus, a piece of glass was instrumental in the birth of modern atomic and nuclear science. The favor was returned, in a sense, in the 1920's when X-ray diffraction studies laid the basis for the modern ideas of glass structure.

Studies of these new radiations soon showed their absorption to be a function mainly of the density of the absorber, lead being the most effective of commonly available materials. Lead-containing glasses were already well-known; so a means was at hand for enhancing the radiation absorption of common transparent glasses. The need for protective glass became significant in the 1940's when nuclear reactions first produced large amounts of radio-isotopes, with their chemical processing behind transparent shields. It grew again with the advent of color television in the 1960's, when the high tube voltages of these sets produced significantly more X rays than black and white models. Incorporating lead has become a major means for enhancing the nuclear and radiation utility of glasses.[1-3] This section will update this development, covering mainly the basis of attenuation and the major applications for lead in TV glasses and shielding window glasses.

X-Ray and Gamma-Ray Attenuation Glasses

Basic Attenuation

Since the thickness of common glass articles is sufficient to absorb alpha and most beta particles, this discussion will cover only the more penetrating X and gamma radiation, where special glasses are needed. Radiation from about 100 Å (10 nm) (0.12 keV)* to 0.1 Å (0.01 nm) (124 keV) are usually called X rays. Shorter wavelengths/higher energies are termed gamma rays, although the overlap is not rigorous.

The attenuation of a narrow, monoenergetic beam is given by a Beer–Lambert type expression

$$I/I_0 = \exp[-\mu x] \tag{1}$$

where I/I_0 is the fractional transmission, μ the linear attenuation coefficient of the material,[†] and x the absorber thickness. Frequently a mass attenuation coefficient (μ/ρ) is calculated, where ρ is the density of the material, and Eq. (1)

---

*[Energy (keV)]/[wavelength (Å)] = 12.398.

[†]Removal from the beam is "attenuation"; the transfer of energy to the medium is termed "absorption."

becomes

$$I/I_0 = \exp[-(\mu/\rho)\rho x] \tag{1a}$$

Mass attenuation coefficients are used because the value of this coefficient for a mixture is readily calculated from the composition as

$$(\mu/\rho)_{mix} = \sum(\mu/\rho)_i W_i \tag{2}$$

where $(\mu/\rho)_i$ is the value of the component i and W_i is the weight fraction. The components are usually taken as chemical elements, but oxide compositions may be used if the μ/ρ values of each oxide are first computed from the elements in the oxide. Once μ/ρ is known for the mixture, it is common to multiply by the density, ρ, of the mixture to get the μ of the mixture and use Eq. (1) for final calculations. The common units for μ are cm^{-1} and for μ/ρ are cm^2/g.

Attenuation coefficients for the elements are tabulated as a function of X- and gamma-ray energy. A prime source for data has been the National Bureau of Standards, Center for Radiation Research.[4-6] Other compilations are also available,[7-9] including values calculated for the oxides in glasses and ceramics[10-12] (Table I).

Table I. Approximate Compositions of Solder Glasses

Oxide	Glass code				
	7570	7520	7578	7574	7594
SiO_2	3.5	71	5	12.5	7.5
B_2O_3	11.5	11.5	9	22.5	7.5
Al_2O_3	10.5	5	1		
PbO	74.5		60		68
Na_2O		10.5			
NaCl		2.0			
ZnO			25	65	5
TiO_2					12

	7595	7063	8870	1826
SiO_2	3.0	32.5	34	38
B_2O_3	9	21.5		79
Al_2O_3		10		5
PbO	77.5	36	58.5	37
ZnO	10.0			
BaO	0.25			
MgO	0.25			
F	0.5			
K_2O			6.5	
Sb_2O_3			1	
Li_2O				1

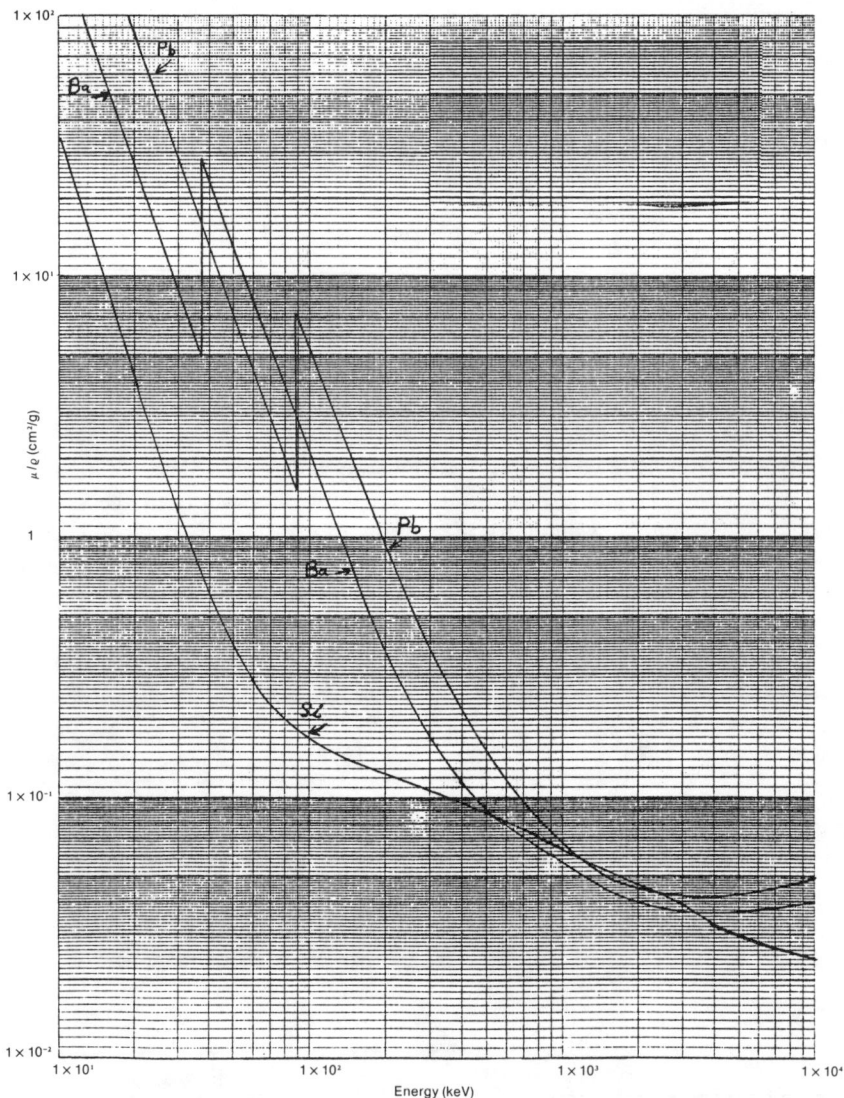

Fig. 1.   Photon mass attenuation in lead, barium, and silicon as a function of energy.

Figure 1 shows the mass attenuation coefficient of a low (Si), medium (Ba), and high (Pb) atomic number (Z) element common to glass and ceramics for 0.10 keV–10 MeV photons. The curve illustrates the rapid decrease in attenuation with energy, up to about 5 MeV, after which the attenuation slowly increases with energy. At lower energies the attenuation is strongly proportional to the atomic number of the elements but is only mildly so at higher energies. Equation (1a) shows that the $(\mu/\rho)\rho x$ product is important. The first two factors depend on composition, but the first depends on photon energy as well. Figure 1 shows that the μ/ρ of a mixture will be most sensitive to composition at low energies but not sensitive at higher energies, especially the region near 0.8–3 MeV. The result is that high Z elements, like lead, *always* enhance attenuation because of their increased density and *sometimes* have a much greater enhancement on μ/ρ, depending on the energy region of interest.

Figure 1 also shows the "absorption edge effect," a sharp increase in the μ/ρ values at atomic energy levels. The practical effect of this is that in narrow X-ray energy ranges a lower Z element may have a μ/ρ value disproportionately large to the effect of that element on the density of a mixture. Since X rays from color TV sets fall into the K edge region of mid-range elements, barium and strontium are useful to enhance the attenuation of TV glasses while minimizing density (hence, weight) increases. Full advantage of this effect is not taken, however, since these elements (or their oxides) are not as compatible as PbO in the glass compositions. This illustrates the point that the usefulness of lead for radiation shielding in ceramic materials is a *combination* of nuclear effects and a favorable ceramic chemistry. For radiation shielding windows, the important gamma-ray energies are in the 1–10-MeV range, beyond absorption edge effects, and in the region where lead plays its key role by increasing the density factor.

Geometry, Dosimetry, and Buildup Aspects

Equations (1) and (1a) assume a narrow beam of radiation perpendicular to an absorber of uniform thickness x. The deviations in real products naturally will modify the simple calculation of attenuation. A TV tube, for instance, is a complex-shaped envelope of sealed pieces of different compositions and thickness, illustrated in Fig. 8 of Chapter 3, part A. The source of X rays is not a simple point or narrow beam. Also, Eqs. (1) and (1a) deal basically with the *number* of particles but not their biological effects. Radiation damage is a combination of the numbers of particles, their types and energies, and the nature of the absorber. Complex medical considerations are involved if the potential absorber is a person. Because of these complications and in order to meet regulations, a measurement of absorbed dose is ultimately needed.[13,14] Attenuation calculations are most useful then to predict the *relative* effects of materials and designs, prior to such confirming measurements of dose.

Another more calculable geometry effect comes up in shielding windows. If the dimension of the radiation beam is large compared to the detector, radiation passing through the edges of the shield in directions not in line with the detector can be scattered toward the detector and enhance the transmission above that predicted by Eq. (1). This is called "buildup" and modifies the point source case as

$$I/I_0 = B \exp(-\mu x) \tag{3}$$

where B, the buildup factor, depends on the energy of the radiation, the Z of the absorber elements, the value of μx, and specific experimental details.[15] The practical effect is that shielding windows must be thicker than predicted by Eq. (1), by an amount comparable to the value of B in Eq. (3), to handle the variety of situations encountered in practice.

TV Glasses

The details of X-ray emission from TV sets and approaches to its measurement and control are well documented in two symposia.[16,17] Industry practices and standards in the U.S. are published by the Joint Electron Device Engineering Council (JEDEC) of the Electronic Industries Association (EIA).[18] The current emission standard is a dose rate of 0.5 mR/h (milliroentgen per hour) measured 5 cm from the tube face. In the glass TV bulb, the faceplate, the funnel, and the neck have different thickness, for mechanical reasons, and are different in composition, densities, and μ values. The development of glass compositions for TV tubes has been reviewed by Sheldon.[19] Lead is an important ingredient in these glasses, especially in the neck and funnel pieces, which are thinner than the faceplate, and thus need a higher μ value to achieve a given attenuation, according to Eq. (1). The large volume production of TV tubes represents the single largest usage of lead for radiation protection.

As shown in Eq. (1a), a low fractional transmission can be achieved by a high μ/ρ value, a high density, a large thickness, or their product. From a manufacturing standpoint, the thickness tends to be fixed by the strength requirements of the various pieces. There are practical limits on the overall weight of the pieces. This limits the densities, leaving the μ/ρ value as the parameter with the greatest range of adjustability, as discussed earlier.

Table II gives data on glass types widely used in color TV bulbs. The industry convention is to compare glasses at 0.6 Å (0.06 nm) (20.7 keV). Although the faceplate glass has the lowest attenuation coefficient, the added thickness (μx product) leads to the largest net attenuation (smallest fractional transmission). The attenuation coefficients of the glasses currently in use have nearly doubled from early compositions, as required by the greater X-ray production from increasing tube voltages.

It should be pointed out that PbO is used in the glasses for one other major reason besides X-ray attenuation, that being the greatly enhanced electrical resistivity, needed for vacuum electronic envelopes, when PbO is substituted for Na_2O, the principal flux in common container and window glasses. Another important property needed in TV glasses is resistance to radiation darkening caused by electrons and X-ray irradiation producing excited states, stable in the glass structure, which absorb in the visible range.

This is a problem only for the faceplate of the tube. The common inhibitor is added CeO_2, but a second-order improvement is achieved by the PbO for Na_2O flux substitution mentioned above, since radiation browning is related to alkali content.

Shielding Window Glasses

TV sets are mass produced, and their radiation levels are moderate and predictable. Radiation shielding windows, on the other hand, are custom-designed assemblies, which must withstand much higher levels of radiation and must serve a wide range of nuclear operations. To put things in perspective, a color TV faceplate can be about 0.5 in. (1.3 cm) thick, 20 in. (51 cm) square, and

Table II. Attenuation of Glasses Used in Color TV Tubes and Radiation Shielding Windows

Glass code	Type	PbO (wt%) (nominal)	μ^* (cm^{-1})	ρ (g/cm^3)	μ/ρ^* (cm^2/g)	Thickness (cm)	Fractional transmission[†]
9068	TV faceplate	2.25	28	2.695	10.4	1.14	1.37×10^{-14}
0138	TV funnel	22.5	62	3.02	20.5	0.25	1.86×10^{-7}
0137	TV neck	28.4	90	3.18	28.3	0.25	1.69×10^{-10}
8365	Shielding	0	0.151	2.70	0.0558	2.2	7.17×10^{-1}
8459	Shielding	35.0	0.187	3.30	0.0568	19.8	2.47×10^{-2}
8455	Shielding	36.0	0.188	3.30	0.0569	19.8	2.42×10^{-2}
8463	Shielding	82.0	0.355	6.20	0.0572	26.5	8.21×10^{-5}
Assembly in Fig. 2							1.48×10^{-11}

*At 0.6 Å/20.7 keV for TV glasses and 1.25 MeV for shielding glasses.
[†]Value of $\exp(-\mu x)$, in Eq. (1), buildup neglected.

weigh about 25 lbs (11.3 kg). The high-density member of a shielding window assembly of 3–5 pieces can commonly be 12 in. (30.5 cm) thick, 30 in. (76 cm) square, and weigh about 2500 lbs (1134 kg); some pieces have been 3 times as thick. Since the window assembly is the transparent section of a hot cell wall, it must shield personnel at least as well as the wall. This implies a shielding capability equivalent to about 4 ft (1.22 m) of high-density (~4 g/cm^3) concrete. Also, the rapid drop in attenuation coefficient with increasing photon energy (see Fig. 1) requires much more glass thickness to achieve a given fractional attenuation in the 1–5-MeV range that hot cells must handle as compared to the 20–30-keV range of TV sets.

As is often the case, such formidable engineering and manufacturing requirements are expensive but not widely needed. Thus, the total usage of lead in shielding glasses is small compared to its application in TV.[20,21]

Figure 2 illustrates a typical shielding window assembly. Some properties of the glasses in this assembly are also given in Table I. Attenuation for shielding glasses is often compared at 1.25 MeV, the average of the two gamma rays of ^{60}Co (1.17 and 1.33 MeV), a common isotope for high dosage radiation processing. As discussed earlier, the reduction in μ/ρ values and a tendency toward similar values for the elements as photon energies increase leaves density as the main factor in attenuation by materials in the 0.8–3-MeV range. This is seen in the similarity of the μ/ρ values in Table I at 1.25 MeV as compared to 20.7 keV (0.6 Å, or 0.06 nm). For the TV glasses, the μ values are much larger and differ mainly due to the effects of composition on μ/ρ values at X-ray energies. For the shielding window glasses at 1.25 MeV, the μ values are much lower and differ mainly due to the effects of composition on density. The fractional transmissions of the glass sections in Fig. 2 are also given, neglecting buildup. The overall transmission fraction of the window assembly, the product of the individual sections, is seen to be higher than that of two of the TV glass pieces.

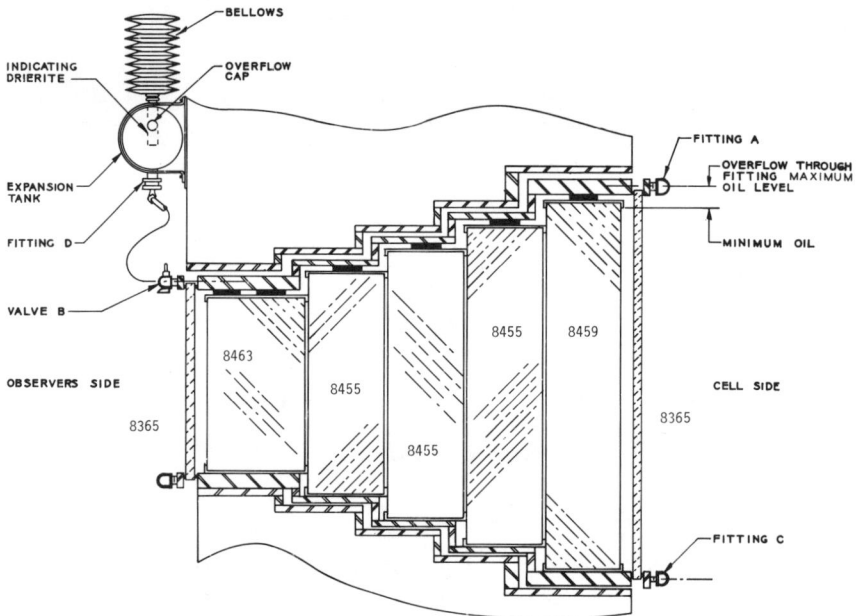

Fig. 2.   Typical shielding window assembly.

Although initial intensities are not comparable, it is interesting to note that a 0.5-in. (1.3 cm) TV faceplate is a better attenuator for X rays than a nearly 4-ft (1.2 m) assembly of shielding glass is for ^{60}Co gamma rays. Functionally, a TV set is a better attenuator than a hot cell!

As with TV glasses, there are other factors besides attenuation involved in shielding windows. Figure 2 shows a chamber around and between the glass segments, which must be very carefully polished. The oil is to reduce reflection losses for viewing. Many windows have this feature, although flexible polymer sheets and optical coatings have been used to reduce the maintenance problems of oil systems. Another factor is radiation browning resistance, which is more critical than in TV glasses because of the viewing differences. One might expect that the densest and better absorbing section of a window assembly would be on the "hot" side of the cell. The opposite is true, as in Fig. 2, because to a crude approximation, the resistance of glass to radiation browning is directly proportional to the silica content while the density varies inversely. Although CeO_2 additions enhance browning resistance, CeO_2-protected high-lead glasses are not as resistant as CeO_2-protected high-silica glasses. The approach is thus to reduce radiation levels several orders of magnitude so that the fullest shielding advantages can be taken of high-lead glasses.

One other unique requirement of shielding glasses has to do with electrical properties. TV glasses, as mentioned, are good insulators. Shielding window glasses must have a certain amount of ionic conductivity to bleed off the charge stored if an intense ionization is delivered over a short time. Otherwise, a spontaneous discharge can occur, a sort of internal "lightning bolt" that has

Fig. 3. Photograph of a spontaneously discharged shielding glass.

caused a dramatic fracture of a few early shielding windows before the effect was understood and controlled by proper glass conductivity.[22] Figure 3 shows a picture, made by using the light of the discharge, of a spontaneously discharged shielding glass.

It should be mentioned that lead has been used for radiation shielding in some ceramic products other than transparent glasses.[23] These include lead-filled concrete, PbO-containing ceramic wall tiles, and lead asbestos composites — all of which have low usage.

Lead in Radiation Detection Glasses

Novel technical applications for high lead glasses depend on their high density/high attenuation, but for detection rather than protection purposes.

The Cerenkov detector, used in high-energy nuclear research, detects the "shock wave" light emitted when a high-energy (100 MeV) particle travels in a medium faster than the phase velocity of light.[24] High-lead glasses are ideal for this application, because they are dense enough for the total absorption of a particle's energy, they can be made in large, excellent optical-quality castings, and they have a high index of refraction (low phase velocity) that enhances the light emission. Another detection application depends on the reduction of PbO to Pb metal by hydrogen on the inner walls of stacked lead glass tubes.[25] The conducting path provided allowed the measurement of the conversion electron current produced by a gamma field.

Summary

Lead is well-known as a major element for radiation attenuation. This is due to both a specific atomic effect, measured by the mass attenuation coefficient, as well as the high density of lead materials. Since lead, as the oxide or silicate, can be readily incorporated in glasses and ceramics, technologists have made wide use of lead to enhance the radiation utility of these materials.

References

[1]R. E. Bastick, "The Uses of Glass in the Field of Atomic Energy," *J. Soc. Glass Technol.,* **42**, 70T (1958).

[2]N. J. Kreidl and J. R. Hensler, "Special Glasses for Nuclear Engineering Applications in Modern Materials," Advances In Development and Applications, Vol. 1. Edited by H. H. Hauser. Academic Press, New York, 1958.

[3]N. J. Kreidl, "Effects of High Energy Radiation"; in Handbook of Glass Manufacture, Vol. II, 1st ed. Edited by F. V. Tooly. Ogden Publishing Co., New York, 1960.

[4]J. H. Hubbell, H. A. Gimm, and I. Overbo, *J. Phys. Chem. Ref. Data,* **9** [4] 1024–148 (1980).

[5]J. H. Hubbell and I. Oberbo, *J. Phys. Chem. Ref. Data,* **8** [1] 69–105 (1979).

[6]J. H. Hubbell, "Photon Cross-Sections, Attenuations, Coefficients and Energy Absorption Coefficients From 10 KeV to 100 GeV," *Natl. Stand. Ref. Data Ser. (U.S. Natl. Bur. Stand.)* **NSRDS-NBS29** (1969).

[7]Handbook of Chemistry & Physics. Edited by R. C. Weast. The Chemical Rubber Co.; Cleveland, OH, 1982.

[8]E. Storm, and H. I. Israel, "Photon Cross-Sections From 0.001 to 100 MeV for Elements 1 through 100," LA-3753 (1967).

[9]R. G. Jaeger et al.; Engineering Compendium on Radiation Shielding, Vols. 1 and 2. Springer-Verlag, New York, 1968, 1975.

[10]L. L. Sun and K. H. Sun, "X-Ray Absorbing and Transmitting Glasses," *Glass Ind.,* **27**, 686 (1948).

[11]G. F. Brewster, "Calculated X-Ray Mass Absorption Coefficients of Glass Components," *J. Am. Ceram. Soc.,* **35** [8] 194 (1952).

[12]D. A. Richardson, "X-Ray and α-Ray Absorption Coefficients of A Number of Glasses," *Br. J. Appl. Phys.,* **8**, 11 (1957).

[13]G. J. Hine, and G. L. Brownell; Radiation Dosimetry. Academic Press, New York, 1956.

[14]K. Z. Morgan; Principle of Radiation Protection. Kreiger Publishing Co., Inc., Milbourne, FL, 1973.

[15]A. B. Chilton; Broad Beam Attenuation, Vol. 9; Vol. 1 1968.

[16]Conference on Detection and Measurement of X-Radiation From Color Television Receivers. U.S. Department of Health, Education and Welfare, Washington, DC, 1968.

[17]Radiological Protection Problems Associated with Parasitic X-Ray Emission From Electronic Products; Euratom Symposium, Toulouse, 1970. C.I.D., Euratom, Luxembourg, 1971.

[18]Recommended Practice for Measurement of X-Ray Radiation From Direct View TV Picture Tubes, JEDEC Pub. No. 641; Considerations Used In Establishing the X-Radiation Ratings of Monochrome and Color Picture Tubes, JEDEC Pub. No. 94. Electronic Industries Association, Washington, DC, 1975.

[19]J. L. Sheldon, Progress In The Development of X-Ray Absorbing Glasses For Television Picture Tubes, Ref. 17.

[20]A. Jacobson and W. Jahn, "The Transparent Lead Wall," *Glass,* **56** [2] 67 (1979).

[21]W. Jahn, "Shielding Materials. Materials Against Gamma Rays. Transparent Shielding Materials. Silicate and Lead Glasses"; in Ref. 9, Vol. 2, 1975.

[22]F. C. Hardtke and K. R. Ferguson, "The Fracture by Electrical Discharge of α-irradicated Shielded Window Glasses"; p. 369 in Proceedings of the Conference on Hot Laboratory Equipment, 1963.

[23]V. Lach, "Shielding Materials. Materials Against Gamma Rays. Ceramics"; in Ref. 9, Vol. 2, 1975.

[24]W. J. Price; Nuclear Radiation Detection. McGraw-Hill Book Co., New York, 1958.

[25]G. K. Lum, M. I. Green, V. Perez-Mendez, and K. C. Tam, "Lead Oxide Glass Tubing Converters for Detection in MWPC;" *IEEE Trans. Nucl. Sci.,* **NS27** [1], 157 (1980).

Fiber Optics

One of the fastest growing technologies today is fiber optics. While normally associated with long-length communications, these high efficiency light pipes find usefulness as faceplates for cathode-ray tubes, bundles for remote illumination, rigid and flexible coherent image transport devices, and decorative lamp components — to name a very few. Lead oxide constitutes one of the major ingredients in most of the applications today. The only exception at the present time is high-efficiency, long-haul communication fiber. Here, fused silica, boria, and germania are used exclusively for purity reasons.

The basic optical fiber is a simple component, consisting of a high refractive index core material and a lower index cladding member (Table I). The ability of an optical fiber to guide light from end to end rests solely on the total internal reflection phenomenon. (The index difference between core and clad establishes a numerical aperture (NA) and, thus, the light gathering and maximum transport capability.) The actual amount of light energy that a fiber transmits is determined by a variety of considerations which include impurity absorptions, scattering losses, bending losses, and other defect-related phenomena. Collectively, these latter considerations are grouped under the heading of fiber attenuation.

There are three quality levels generally associated with the fiber attenuation parameter. They are high loss (400–1000 dB/km), medium loss (50–400 dB/km), and low loss (less than 50 dB/km). A fourth category, ultra-low loss, should be included in modern literature to describe "state-of-the-art" fiber which exhibits less than 1 dB/km (1 dB or decibel of attenuation is defined as 10 times the $\log_{10}$ of the input/output power ratio for the fiber). Fiber attenuation is wavelength dependent and generally less in the near-infrared spectrum (700–1600 nm) than in the visible portion (400–700 nm).

High-loss fibers find application in devices where the optical path length of the component is very small and normally less than 10 mm. Generally, they are used in the visible portion of the spectrum, where typical applications include computer card readers, LED transmitters, and imaging faceplates. A 1000-dB/km fiber still transmits over 90% of the incident energy at a 10-mm thickness.

Medium-loss fibers are used for remote location illuminators, short-haul computer data links, optical thyristor switching, and flexible image bundles. The former three components are used with either visible or near-IR light. The latter device uses very fine fibers (about 15 μm in diameter) to map out a coherent image from one face plane to the other and is used almost exclusively in the visible portion of the spectrum. Typical lengths for all of these items range between 2 and 10 m. At 10 m, a 400-dB/km fiber transmits about 40% of its incident light energy.

Low- and ultralow-loss fibers are used in the fast-growing communication industry (voice and data) to replace expensive copper wire and provide enhanced channel multiplexing capability. Without exception, the application wavelengths all lie in the near-IR portion of the spectrum, where intrinsic and extrinsic losses are inherently lowest. A 0.5-dB/km fiber operating at 1550-nm wavelength can

Table I.   Lead Core Glasses for Use with Soda-Lime Silicate Cladding Glass in the Manufacture of Optical Fiber

index (n_D)	NA	PbO	SiO_2	B_2O_3	BaO	ZnO	Al_2O_3	Na_2O	K_2O	As_2O_3
1.545	0.31	16.3	62.2	2.5		1.2		6.0	11.6	0.2
1.581	0.46	34.2	52.4					6.3	6.9	0.2
1.623	0.59	46.5	44.8				2.0		6.5	0.2
1.725	0.83	41.0	16.5	14.5	8.2	14.0	5.5			0.3
1.808	0.99	67.1	29.0				1.5	1.0	1.0	0.4

(Identification columns: index (n_D), NA. Composition (wt%): PbO, SiO_2, B_2O_3, BaO, ZnO, Al_2O_3, Na_2O, K_2O, As_2O_3.)

transmit digital data over 40 km without the need for interruption and signal amplification. Such a fiber currently costs about $2–3 m^{-1}. Even a relatively inexpensive ($0.50 m^{-1}) 3-dB/km, low-loss fiber is useful to distances over 5 km.

Lead oxide containing glass plays a vital role in all but the lowest loss fibers. The major reasons for this are as follows: (1) the high refractive index contribution permits the manufacture of numerical apertures approaching 1.0; (2) most of the transition-metal elements found as impurity absorbers exhibit a greatly diminished effect in lead-based glasses; (3) a wide range of viscosity and thermal expansion properties are obtainable to improve the fiber strength characteristics; (4) suitable high-purity starting materials such as Hammond UHP red lead are readily and inexpensively available.

The intensity and spectral distribution of the impurity absorption bands for the transition-metal ions are dependent on the glass matrix. In lead glass, the elements iron (Fe), copper (Cu), and chromium (Cr) produce much smaller extinction coefficients than they do in silicate and borosilicate bases. As a result, higher concentrations of these impurities are permitted in leaded fiber optics. Hammond code UHP red lead oxide typically contains less than 8 ppm Fe, 2 ppm Cu, and 0.8 ppm Cr. These impurity levels are consistent with the remainder of the glass composition oxides chosen and yield core materials which exhibit average visible attenuations of less than 250 dB/km.

At the 150-dB/km level and a PbO cost of less than $5.00/kg, it becomes commercially feasible to manufacture devices which are presently cost prohibitive (i.e., medium-length, disposable image scopes for underwater periscopes and nuclear reactor applications). This attenuation goal is readily achievable with Fe and Cu levels both below 2 ppm. At the 50-dB/km level and a PbO cost of less than $200/kg, it becomes commercially feasible to manufacture long-length devices of this nature (e.g., about 100 m). This goal will require a material with Fe, Cu, and Cr impurity levels all below 0.8 ppm.

In conclusion, the world of fiber optics relies heavily on the use of lead oxide containing glasses. Many suitable glass compositions are currently available and in use today. Many more would be developed and placed into high-volume applications with the advent of relatively low-cost, raw material purification improvements.

Prereaction in Lead Glass Melting

Prereaction of lead glass batches can reduce raw material dusting and batch segregation. Reduction of lead volatility in glassmaking is also very important, and use of lead silicates has helped to achieve this. Ott et al.[1-7] further demonstrated in small crucible studies of melting reactions and kinetics that volatility can also be significantly reduced by prereaction of batch materials in the solid state. Later it was demonstrated by Ganzala et al.[8-9] with a model glass melter that volatility reduction by batch prereaction could be expected in the continuous glassmaking process. This section on prereaction will summarize what was found by Ott and Ganzala.

The batch used in these studies was representative of a medium lead tubing composition. The batch composition is shown in the following table:

Component	Composition (wt%)
SiO_2	49.47
$3PbO\text{-}SiO_2$	30.99
K_2CO_3	11.85
Na_2CO_3	5.56
Al_2O_3	1.20
$NaNO_3$	0.47
Sb_2O_3	0.47

In the small crucible studies, combinations of the batch components were mixed and analyzed by differential thermal analysis (DTA), thermogravimetry (TG), differential thermogravimetry (DTG), and X-ray study. From this study the fundamental melting reactions and associated kinetics were better understood.

Evaluation of the four-component, silica–tribasic lead silicate–sodium carbonate–potassium carbonate, system resulted in two major conclusions. The first was that the cause of segregation was the low melting point and high fluidity of the tribasic lead silicate. The second was that solid-state reaction was observed with fine materials at temperatures below that at which significant volatilization of the lead oxide occurs. These results indicated the potential benefits of solid-state prereaction.

Prereaction in the solid state should reduce dusting by forming a solid clinker. Segregation should be reduced because the fluidity during melting would now be controllable. Would lead volatility, however, actually be reduced under the dynamic conditions of continuous glassmaking? This question was explored with the model continuous melter. In another lead glass system, Preston and Turner[10] had reported that there was significant volatility during the early stages of melting.

Before discussing the model melter experiments, the fundamental reactions in the system studied here need to be reviewed. In the two-component system, silica–ground tribasic lead silicate, reaction began at about 650°C with the formation of several silicate compounds. These silicate compounds melted eutectically and gradually formed more liquid lead silicate. This reached a maximum at about 763°C, and further heating served only to dissolve the remainder of the silica.

When the soda ash and potash were added to form a four-component system, a sodium–lead–silicate was formed at 600°C. Over the range 650–715°C, the silica and tribasic lead silicate reacted. Up to about 740°C the alkali carbonates each reacted independently in the silica–tribasic lead silicate matrix. The major decomposition occurred from 770 to 830°C. Increasing temperature facilitated solution of the remaining silica by the reactive liquid phase.

When the alumina was added, the temperature of the reactions was not affected, nor was the sequence of the reactions affected. The addition of the two fining agents, sodium nitrate and antimony trioxide, produced a definite effect; the reactions proceeded more rapidly. The sequence of the reactions, however, was unaltered.

It was also found that the volatility of the lead was not inherent in the actual melting process but rather was the result of segregation either during melting, with the lead oxide chemically uncombined or poorly combined and, hence, more easily volatilized, or during the dusting of fine materials.

Litharge, or lead monoxide, is a primary material used for introducing lead into glass batches. Litharge, however, is subject to dusting due to its finely divided form. A way to circumvent this problem is to convert it to lead silicate by reaction with silica. From a review of lead oxide sources, it was concluded that the nature of the lead oxide addition affects both the volatility of the batch materials and segregation during melting of the batch. On the basis of this review and the small crucible research, it was concluded that a prereacted material, containing all the necessary ingredients, would be the best form to make the lead and alkali additions.

Batches for the model continuous melter studies were prereacted at low temperature (750°C) in the solid state in a commercial tunnel kiln. The prereacted batches used either four (alumina and fining agents missing) or five (only fining agents missing) of the seven components. The prereacted clinker was then crushed, mixed with the remainder of the components, and then melted for several days in the model tank. The same composition, not prereacted, was also melted for comparison.

During melting, samples were drawn periodically from the tank. Density was measured after annealing and subsequently converted to percent lead loss from the batch. Before preparation of any one prereacted batch for continuous melting, statistical optimization based on thermogravimetry was conducted to ensure optimum prereaction. Thermogravimetry was also used to establish the linear relationship between glass density and percent lead loss due to volatility.

Lead volatility of the four- and five-component prereacted batches was about half the volatility of the commercial batch. Under all test conditions neither of the two prereacted batch compositions experienced the segregation of the commercial batch. Also, although the prereacted batches were formulated with fine materials (silica and tribasic lead silicate, -325 mesh), they did emerge from the tunnel kiln as a hard sintered product that easily crushed to -10 mesh. Therefore, the prereacted batches not only reduced lead volatility and segregation, but also demonstrated the potential to reduce dusting.

References

[1]W. R. Ott, "Reaction Kinetics in Lead Based Glass Melts"; Ph.D. Thesis, Rutgers University, New Brunswick, NJ, 1970.

[2]W. R. Ott, "Subsolidus Studies in the System $PbO-SiO_2$," *J. Am. Ceram. Soc.*, **53** [7] 374–5 (1970).

[3]W. R. Ott and M. G. McLaren, "DTA/TGA Investigation of the Reaction in Sodium-Lead-Silicate Glass Batch"; pp. 1329–45 in Thermal Analysis, Vol. 2. Academic Press, New York, 1969.

[4]W. R. Ott, "Simultaneous TG/DTG/DTA Evaluation of a Lead Glass Batch"; Mettler Thermal Technique Series, Technical Bulletin T-108, 1970.

[5]W. R. Ott, "Kinetics and Mechanism of the Reaction Between Lead Orthosilicate and Potassium Carbonate"; pp. 579–90 in Thermal Analysis, Vol. 3. Edited by Wiedemann. Birkhauser Verlag, Basel, Germany, 1972.

[6]W. R. Ott, M. G. McLaren, and W. B. Harsell, "Thermal Analysis of Lead Glass Batch," *Glass Technol,* in press.

[7]W. R. Ott, B. K. Speronello, and C. J. Brinker, "Lead Silicate–Potassium Carbonate Solid-Solid Reaction Kinetics," *Thermochim. Acta,* **6**, 85–94 (1973).

[8]G. W. Ganzala, "Development of a Prereacted Batch Material for a Medium Lead Content Glass Using a Model Continuous Glass Tank"; Ph.D. Thesis, Rutgers University, New Brunswick, NJ, 1973.

[9]G. W. Ganzala and W. R. Ott, "Studying Prereacted Batches Using a Model Continuous Glass Tank," *Glass Ind.,* **54** [6] 20–21 (1973).

[10]E. Preston and W. E. S. Turner, "A Study of the Volatilization and Vapor Tension at High Temperature of an Alkali-Lead Oxide-Silica Glass," *J. Soc. Glass Technol.,* **16**, 219 (1932).

4

Ceramic Glazes

Sweetener! That is what some ceramists call lead. Even small amounts of lead can dramatically change a hard, unforgiving glaze to one that flows smoother and is temperature tolerant. In fact, lead is considered essential for applications where the glaze must be brilliant, of high luster, and of such smoothness that it is absolutely free of defects.[1] Substitutes for lead in such applications fall far short of current manufacturing requirements.

Over the centuries, a number of lead-bearing compounds have been used in ceramic glazes. Galena (PbS) was used as a glaze material by prehistoric Chinese craftsmen.[2] Until recent times, it was also used in the production of the cruder forms of pottery in Great Britain and the U.S. Often dry galena was simply dusted on wet ware before glost firing.[3] However, white lead, $2PbCO_3 \cdot Pb(OH)_2$, because of its high purity, eventually became the preferred material for potters.

White lead, in contrast to litharge (PbO) and red lead (Pb_3O_4), has small particle size and lower particle density, which makes it capable of suspending a glaze without the presence of clays.[4] Unfortunately, white lead (along with the oxides of lead) is more soluble than galena and is being phased out of use by the ceramic industry.

Modern practice is to add lead to glaze in frit form. A frit is a glass prepared according to a set of principles called the fritting rules.[5] The purpose of fritting is to desolubilize and detoxify specific chemical elements. It also permits lower glaze-firing temperatures and improves uniformity. Frits are generally produced by frit manufacturers rather than ceramic manufacturers. Each frit has an identification number and usually a published composition. Some frits are specifically produced for various ceramic companies, and their compositions are considered confidential or proprietary by the frit supplier. Chapter 7 discusses the chemistry of frits in detail.

Table I shows typical published data. When the frit composition is known, it can be used as any other compound to formulate glazes of known composition according to standard ceramic procedures.[6] Two less expensive frits, "lead monosilicate" and "lead bisilicate," have been useful in the past in that they reduce dust generation during glaze batching. Special stabilized variations of these frits have significantly lower solubilities than white lead.

Alumina (Al_2O_3) has been found to be an excellent stabilizer for lead silicate frits. Titania (TiO_2) is also very effective, although color can be a problem.[7] Frit suppliers continue to develop frits having still lower solubilities. Frit solubility should be carefully considered in all new glaze formulations.

Table I. Commercial Lead Frit Properties

Frit	Approx. exp. coeff. ($\times 10^{6\circ}$C)	Melting range (°F)	Approximate formula*		
30% PbO	6.8	1350–1460	$0.21Na_2O$ $0.29CaO$ $0.50PbO$	$0.12Al_2O_3$ $0.66B_2O_3$	$2.6SiO_2$ $0.7ZrO_2$
17% PbO	6.5	1590–1670	$0.09K_2O$ $0.09Na_2O$ $0.58CaO$ $0.24PbO$	$0.19Al_2O_3$ $0.36B_2O_3$	$2.8SiO_2$
65% PbO (lead bisilicate)	6.7	1400–1500	$1.0PbO$	$0.11Al_2O_3$	$2.16SiO_2$

*Supplier should be contacted for actual formula.

Lead borosilicate glazes should contain two frits, the lead (Pb) being in one frit and the boron, soda, and potash in the second frit. This will reduce the overall solubility of the unfired glaze.[8]

Flake frit, produced by roller cooling, or granular frit, produced by water quenching, will prevent dusting during milling operations, the flake being the easiest to grind.

The solubility of a frit or glaze will depend on its particle size and the solubility of its individual components. Solubilities are measured in terms of the amount of dissolved PbO generated during the particular solubility test. Both the Currier method and the BCRA method are used.[9] The Currier method utilizes a 1-h leach, in 0.14N HCl at a temperature of 40°C. The BCRA (British Ceramic Research Association) method specifies a 2-h leach in 0.07N HCl, the first hour under agitation, at a temperature of 23±2°C. The particle size distribution must be specified to make meaningful comparisons between materials.

The above tests were developed to relate to the human digestive system, but clinical tests have shown that materials of lower solubility are also much less likely to be absorbed on inhalation.[10] In Great Britain raw glazes having a PbO solubility less than 5% (BCRA) are exempt from further regulation other than good housekeeping and regular dust control.[11] Some U.S. potters have requested that OSHA consider such regulations for the U.S. ceramic industry.[12]

Table II illustrates the relative lead (Pb) solubilities of various ceramic glaze materials.

Glaze Preparation and Application

Glaze is almost always applied in liquid form, although there are some exceptions. Fritted glazes usually need the addition of clay or organic binders to assure that the frit stays in suspension and that the glaze does not dust off the ceramic ware to which it is being applied prior to firing. The glaze materials are ball-milled to a particle size distribution that will permit uniform application but not so fine that the glaze lead solubility exceeds standards set for safety reasons.

Table II. Lead (Pb) Solubilities of Various Glaze Materials

Material	Particle size	BCRA	NIOSH
White lead		87–94	81–88
Lead monosilicate		92–94	88–94
Galena		1.6	4.3
Lead bisilicate	50% < 11 μm	8	43
(Al$_2$O$_3$ stabilized)	60	9	47
	70	13	52
	80	17	57
	90	19	62
17% PbO frit	40% < 11 μm	0.47	3.9
	50	0.56	4.7
	60	0.65	5.5
	70	0.74	6.2
	80	0.83	7.0
Improved lead bisilicate	50% < 11 μm	1.4	11
	60	1.9	11.6
	70	2.3	12.5
	80	2.8	14.0

Spraying is probably the most common method of glaze application in the ceramic industry. The glaze may be applied to green ware or to bisque ware according to the particular manufacturing process being used. Bisque ware may be porous or vitreous according to the bisque firing temperature. If the ware is vitreous, it may require heating before glaze application.

Various glaze spray machines have been developed. They may be circular or straight line. They are generally capable of rotating the ware and have multiple spray guns, which can be oriented according to the item being sprayed. Glaze thickness must be controlled and uniform. This is easily done by direct measurement of the thickness and by weight measurements.

Dipping is an ancient process for glaze application. It requires a talented "flick of the wrist" and is generally used only on shapes that are not amenable to spraying. Simple shapes like wall tile can be glazed by a waterfall technique. Lead glazes are generally more tolerant to application variables than nonlead glazes.

Dinnerware

Dinnerware is produced by plastic forming, slip casting, and dry pressing. The composition may be either the traditional triaxial variety or the bone china. The important point is that the glaze and body compositions must be such that the glaze, after firing, is in suitable compression to resist any subsequent tensile forces generated in service. In the case of porous bodies, these forces may be generated by moisture absorption resulting in body expansion.

The thermal expansion coefficient of the glaze should be less than that of the body. It is often easier to adjust the expansion of the body rather than the glaze.[13] Sufficient thermal shock testing, and in the case of porous ware, autoclave testing, should be performed to guarantee the stability of the glaze. All production processes including decorating should be completed before the ware is tested.[14]

Ware that is ready for the consumer must be tested for lead (Pb) release by standard methods specified by the International Standards Organization.[15] It is more likely, if a problem develops, that it is due to decorations rather than the glaze. Such problems are generally easily rectified by firing or compositional changes.

Compositional data for lead-bearing glazes for dinnerware are shown in Table IIIA and Table IIIB. Table IIIC shows various whiteware glaze compositions.

Electrical or Electronic Applications

There are important properties of lead-bearing glazes which make them of particular interest for electronic or electrical applications.[16] In such applications, the bulk and surface properties of the glaze come into play.

Surface resistivity under high humidity conditions is determined as a function of test duration for military application (JAN-1-10). If the glaze is not chemically resistant and if the surface does not remain flaw free throughout the test, it is unlikely that the minimum resistivity of 7.5×10^5 MΩ could be obtained. Such tests usually run for 60 days at a relative humidity of 95%.

Lead glazes are effective in tying up alkali ions, which if released on the surface of a dielectric would decrease the surface resistivity. Also, the lead glazes do not wet easily. This in turn inhibits dissolution of alkalies from the glaze surface.

Table IV shows glazes which may have suitable properties for JAN-1-10 applications.

Structural Clay Products

In contrast to ceramic body stains, lead-bearing glazes can be colored to yield a wide palette of colors for structural products at low cost.[17] Properly formulated glazes show good resistance to weathering against the elements, including "acid rain." These glazes also have good covering power and can readily adapt to the irregular surface of clay products. Brick and tile that would normally be rejected due to chips and rough surface can be reclaimed in glazing operations and then sold at a premium price.

Glaze can be sprayed on unfired clay products, but there are several advantages to using a two-fire process: (1) a wider color palette, (2) ease in reclaiming defective brick, tile, etc., (3) better flow and chemical resistance due to higher lead content, and (4) ease of producing small lots, as often requested by architects; this may include the use of smaller periodic kilns rather than tunnel kilns.

Table IIIA. Average Compositions and Standard Deviation in Composition of Lead Glazes, including Calculated Thermal Expansion Coefficient, by Glaze Maturing Range*

	880–900°C (1616–1652°F)		920–960°C (1688–1760°F)		980–1000°C (1796–1832°F)		1020–1060°C (1868–1940°F)		1120–1140°C (2048–2084°F)		1160–1200°C (2120–2192°F)	
	Normality (N)	Compn. (wt%)	Normality (N)	Compn. (wt%)	Normality (N)	Compn. (wt%)	Normality (N)	Compn. (wt%)	Normality (N)	Compn. (wt%)	Normality (N)	Compn. (wt%)
Na_2O	0.055	1.27	0.088	1.66	0.144	2.54	0.146	2.66	0.146	2.60	0.082	1.49
SD	0.135	3.1	0.118	2.22	0.140	2.31	0.082	1.31	0.122	2.13	0.089	1.54
K_2O	0.017	0.42	0.082	2.38	0.035	0.058	0.161	4.75	0.144	4.02	0.179	4.89
SD	0.041	1.02	0.100	2.95	0.986	1.66	0.054	1.89	0.094	2.54	0.053	1.44
CaO	0.155	2.7	0.223	3.79	0.289	5.16	0.299	5.28	0.370	7.80	0.412	7.18
SD	0.17	3.07	1.69	2.80	0.201	3.71	0.115	2.44	0.157	5.90	0.076	1.73
ZnO			0.013	0.350	0.009	0.243					0.053	1.41
SD			0.043	1.21	0.035	0.909					0.068	1.82
PbO	0.772	47.7	0.595	37.8	0.530	32.3	0.395	27.4	0.373	24.8	0.282	19.2
SD	0.255	12.5	0.287	16.0	0.240	15.9	0.152	11.0	0.227	19.4	0.081	5.29
B_2O_3	0.213	4.02	0.262	5.50	0.364	7.38	0.278	5.78	0.290	4.50	0.302	6.40
SD	0.228	6.17	0.270	5.63	0.203	4.07	0.197	3.97	0.228	3.46	0.124	2.43
Al_2O_3	0.205	6.3	0.263	7.60	0.215	6.34	0.230	7.20	0.324	9.62	0.290	8.91
SD	0.134	2.60	0.047	2.00	0.058	1.72	0.079	2.18	0.135	3.17	0.051	0.992
SiO_2	2.12	36.7	2.33	40.7	2.50	43.7	2.54	46.9	2.59	45.6	2.75	50.4
SD	0.341	5.4	0.374	5.27	0.39	6.72	0.464	4.69	0.966	10.4	0.342	1.91
Calcd. therm exp. ($\times 10^7$°C)		98.1		95.8		87.2		94.4		97.1		87.3

*Calculated from lead glaze formulation in Ref. 4.

Table IIIB. Ranges of Compositions of Lead Glazes, including Calculated Thermal Expansions, by Glaze Maturing Range*

	880–900°C (1616–1652°F)		920–960°C (1688–1760°F)		980–1000°C (1796–1832°F)		1020–1060°C (1868–1940°F)		1120–1140°C (2048–2084°F)		1160–1200°C (2120–2192°F)	
Na₂O (N)	0	0.33	0	0.3	0	0.25	0	0.25	0	0.25	0	0.25
(wt%)	0	7.6	0	5.2	0	7.7	0	4.7	0	4.6	0	3.9
K₂O (N)	0	0.10	0	0.25	0	0.15	0.091	0.25	0.05	0.30	0.05	0.25
(wt%)	0	2.50	0	6.7	0	4.4	2.4	7.8	1.5	8.3	1.5	6.0
CaO (N)	0	0.33	0	0.5	0	0.7	0.15	0.5	0.20	0.6	0.25	0.50
(wt%)	0	6.90	0	7.9	0	13.1	2.4	10.3	4.1	18.1	3.5	9.0
ZnO (N)			0	0.15	0	0.13					0	0.145
(wt%)			0	4.2	0	3.4					3.4	3.8
PbO (N)	0.33	1.0	0.25	1.0	0.25	0.9	0.25	0.7	0.20	0.7	0.127	0.45
(wt%)	27.3	61.4	15.7	59.8	16.3	57.7	16.9	45.3	11.0	58.1	14.1	31.8
B₂O₃ (N)	0	0.53	0	0.6	0	0.76	0	0.5	0	0.62	0	0.5
(wt%)	0.9	14.9	0	11.8	0	15.4	0	10.6	0	10.2	0	8.8
Al₂O₃ (N)	0.13	0.40	0.20	0.35	0.1	0.3	0.14	0.3	0.13	0.45	0.25	0.34
(wt%)	4.8	10.5	2.50	9.70	3.4	9.0	4.5	10.8	5.0	13.5	7.8	10.8
SiO₂ (N)	1.73	2.30	1.85	3.00	1.85	3.0	1.64	3.21	1.3	3.93	2.3	3.0
(wt%)	31.0	38.5	30.7	51.1	32.1	54.2	37.4	51.4	29.2	55.9	47.9	53.4
Calcd. therm exp (×10⁻⁷°C)	85.31	111.2	70.3	106.1	71.8	105	81.9	107	84.2	111.4	73.9	97.3

*Calculated from lead glaze formulation in Ref. 4.

Table IIIC. Various Whiteware Glaze Compositions (Ref. 4)

Glaze	Cone	Colorant	Molecular formula		
Red orange	012–08	2% Cr_2O_3	$1PbO$	$0.101Al_2O_3$ $0.038Cr_2O_3$	$0.203SiO_2$
Robin's egg blue	06–05	5% CrO 1.3Fe_2O_3	$0.68PbO$ $0.16CaO$ $0.10K_2O$ $0.06ZnO$	$0.35Al_2O_3$	$2.15SiO_2$ $0.10SnO_2$
Clear lead bright	04	For matte substitute BaO for CaO	$0.6PbO$ $0.3CaO$ $0.1K_2O$	$0.25Al_2O_3$	$1.75SiO_2$
Light blue bright medium opaque	08–1	Copper Oxide & zirconium spinel: 40.3% ZrO_2, 19.6ZnO 20.4SiO_2, 19.7Al_2O_3	$0.448PbO$ $0.174Na_2O$ $0.058K_2O$ $0.186CaO$ $0.080ZnO$ $0.054CuO$	$0.265Al_2O_3$ $0.420B_2O_3$	$2.137SiO_2$ $0.163ZrO_2$
Wall tile glaze	01	Color & opacify as needed	$0.194PbO$ $0.187Na_2O$ $0.192CaO$ $0.051BaO$ $0.376ZnO$	$0.235Al_2O_3$ $0.179B_2O_3$	$1.895SiO_2$
Wall tile glaze	4	Color & opacify as needed	$0.116PbO$ $0.284Na_2O$ $0.287CaO$ $0.015BaO$ $0.298ZnO$	$0.345Al_2O_3$ $0.183B_2O_3$	$2.704SiO_2$
Semivitreous dinnerware	5		$0.26PbO$ $0.43CaO$ $0.12K_2O$ $0.06Na_2O$ $0.13ZnO$	$0.27Al_2O_3$ $0.31B_2O_3$	$2.6SiO_2$
Crystalline glaze	7		$0.50PbO$ $0.20CaO$ $0.09Na_2O$ $0.01K_2O$ $0.20ZnO$	$0.11Al_2O_3$	$2.04SiO_2$ $0.50TiO_2$

Although some color deterioration and microcrazing may be tolerated (or even desired by the architect), it is important to test the final product by autoclave, thermal shock, and chemical methods.[18]

Possible glaze compositions for structural clay products are listed in Table V.

Sanitary Ware

Lead-bearing glazes are used in the sanitary-ware industry to reclaim defective ware at a lower firing than the original glost firing. The reasons for using

Table IV. Glazes for JAN-1–10
Applications (Ref. 16)

Lead monosilicate (wt%)	Additive (wt%)
75–95	5–25kaolin
70–80	20–30Al$_2$O$_3$
95	5Cr$_2$O$_3$
90–95	5–10SiO$_2$
90–99.5	0.5–10SnO$_2$
90	10TiO$_2$
70–99.5	0.5–30ZrO$_2$

Table V. Glazes for Structural Clay Products (Ref. 17)

		Matte (wt%)	Gloss (wt%)
One-fire opaque glaze for	Lead bisilicate	31.6	49.5
red shale brick	Feldspar	16.4	16.2
Cone 06	Flint	6.8	2.9
	Ball clay	13.5	8.6
	Zircon	14.9	13.2
	Wollastonite	6.3	6.2
	Zinc oxide	7.7	3.2
	Alumina hydrate	2.8	
Low-fire clear glaze for	Lead bisilicate	73	
twice-fired brick	Zinc oxide	6	
about Cone 010	China clay	8	
	Zirconia	6	
	Flint	7	

lead-bearing glazes are similar to those already stated for other products. Table VI shows typical glaze compositions for sanitary-ware reclaim.

Formulating Lead-bearing Glazes

Using a particular glaze formulation from the literature may not provide a suitable glaze in practice for the following reasons:

(1) Different body composition, which may have
  (a) a different thermal expansion coefficient
  (b) different surface wetting characteristics
  (c) different porosity or a change of glaze/body interaction chemistry
(2) Different firing conditions such as kiln atmosphere on firing cycle
(3) Different glaze formulation or preparation although the composition is the same
  (a) different ingredients (mineralogical forms and impurities)
  (b) different particle size distribution and/or organic binders, etc.

Table VI. Reclaim Glazes for Sanitary Ware

Cone		Formulation		
3–5		85% Pb-906		
		5% zircon		
		10% kaolin		
	Pb-906 Frit	0.24 PbO	$0.19Al_2O_3$	$2.96SiO_2$
		$0.09 K_2O$	$0.36B_2O_3$	$0.14ZrO_2$
		$0.09 Na_2O$		
		0.58 CaO		
08–1 (Ref. 4)		0.10 PbO	$0.14Al_2O_3$	$2.770SiO_2$
		$0.058 Na_2O$		$0.067SrO_2$
		$0.015 K_2O$		$0.325ZrO_2$
		0.607 CaO		
		0.270 ZnO		
		0.028 BaO		
		0.012 MgO		

(4) Different application methods or skills

Moreover, a glaze which has been developed in the laboratory by a frit supplier for a particular plant application may not function properly in the plant. In this case it is almost always an application or firing problem. In either case, the glaze formulation must be modified to fit the actual intended function.

Adjusting the Thermal Expansion Coefficient

If the thermal expansion of the glaze is too high, the glaze will craze during thermal shock testing[19] or during autoclave testing.[20] Ware that fails under such specified testing should never be placed in the market place.

To lower the thermal expansion of a glaze, one should consider the chemical constituents of the glaze and then replace those elements contributing to the high expansion coefficient with elements that help to lower the expansion. Sometimes the simple addition of silica or china clay will produce the desired result.

If the expansion of a glaze is too low and peeling (shivering) occurs under test or just on standing, then the glaze can be adjusted by adding alkali or other elements contributing to a high thermal expansion coefficient.

Any element added to change glaze expansion should be considered in light of its effect on the lead solubility of the glaze. Boron should be incorporated in a separate frit from the lead-bearing frit. Table VII shows thermal expansion factors for various chemical elements. Table VIII summarizes possible compositional effects on lead solubility.

Adjusting the Firing Range

Lead-bearing glazes generally have a wide firing range. If the glaze needs to be hardened, the simple addition of china clay or silica may suffice. Lead-

Table VII. Thermal Expansion Factors for Chemical Elements in Glazes

| Compd. | Thermal expansion factor* | | | |
	Winkelman & Schott	English & Turner	Mayer & Havas	F. P. Hall
AlF_3			2.47	
Al_2O_3	1.67	0.14	1.67	0.50
As_2O_3	0.67		0.67	
BaO	1.00	1.40	1.00	1.20
B_2O_3	0.03	$-0.66^\dagger$	0.03	0.20
BaO			1.57	
CaF_2			0.83	
CaO	1.67	1.63	1.67	1.50
CdO			0.85	
CeO_2			1.40	
CoO			1.47	
Cr_2O_3			1.70	
CuO			0.73	
Fe_2O_3			1.33	
K_2O	2.83	3.90	2.83	3.00
Li_2O	0.67		0.67	
MgO	0.03	0.45	0.03	0.20
MnO_2			0.73	
Na_3AlF_6			0.85	
NaF			2.47	
Na_2O	3.33	4.32	3.33	3.80
NiO			1.33	
P_2O_5			0.67	
PbO	1.00	1.06	1.40	0.75
Sb_2O_5			1.20	
SiO_2	0.27	0.05	0.27	Variable
SnO_2			0.67	
TiO_2			1.37	
ZnO	0.60	0.70	0.60	1.00
ZrO_2		0.23	0.70	

*Approximate formula for linear coefficient of expansion is

$$\alpha \times 10^7 = \alpha_1\rho_1 + \alpha_2\rho_2 + \alpha_3\rho_3 + \cdots$$

where ρ_1, ρ_2, etc., are percentage by weight of the oxide in the glass and α_1, α_2, etc., are empiric, specific coefficients of the oxides.
†0–12% B_2O_3.

Table VIII. Compositional Effects on Lead Solubility
in Glazes (Ref. 13, p. 560)

Element	Acid resistance	Base resistance
K_2O, Li_2O	Improves if substituted for Na_2O	
CaO, SrO, BaO, MgO, PbO	Improves if substituted for alkalies	
ZnO	Best improvement when substituted for mono- or divalent oxides	
B_2O_3	Gives improvement over alkalies	Reduces if given content is exceeded
TiO_3	Increases resistance	Decreases resistance
ZrO_2	Increases resistance	Increases resistance
Al_2O_3	Increases resistance	Increases resistance
SiO_2	Increases resistance	Increases resistance

bearing frit can be used to soften a glaze. The degree of hardness must be controlled carefully if decorating firings follow the glost firings. New materials received should be tested before they are placed in production. Minor mill formula corrections may be required due to raw material fluctuation.

Adjusting Application Properties

Clays, gums, dispersants, etc., can be varied in a glaze to improve application to the ware. In high-frit glazes, montmorillonite (bentonitic) type clays can be used in small amounts to help suspend the glaze. Natural or synthetic gums can be used in conjunction with the bentonite. Table IX lists certain materials for possible usage in ceramic glazes.

The Future of Lead in Ceramic Glazes

Because of the special properties that lead imparts to ceramic glazes, it will continue to be used in the ceramic industry. A great deal of work needs to be done in the United States to lower the lead solubilities of commercial ceramic frits. When this work is completed, the future use of lead in ceramic glazes should be assured.

Table IX. Materials for Ceramic Glazes* (Ref. 6, pp. 123–125)

Mineral name	Mineral formula	Formula weight (lb)
Albite (soda spar)	$Na_2O \cdot Al_2O_3 \cdot 6SiO_2$	525.1
Alumina	Al_2O_3	101.9
Anatase (see titania)		
Andalusite	Al_2SiO_5	162.3
Anhydrite	$CaSO_4$	136.2
Anorthite	$CaO \cdot Al_2O_3 \cdot 2SiO_2$	278.7
Antimony oxide	Sb_2O_3	291.5
Aragonite (see calcium carbonate)		
Arsenious oxide	As_2O_3	197.8
Barium carbonate	$BaCO_3$	197.4
Barium chloride	$BaCl_2 \cdot 2H_2O$	244.3
Barium chromate	$BaCrO_4$	253.5
Barium hydroxide	$Ba(OH)_2 \cdot 8H_2O$	315.5
Barium oxide	BaO	153.4
Barium sulfate (barite)	$BaSO_4$	233.4
Bismuth oxide	Bi_2O_3	466.0
Bone ash	$13CaO \cdot 4P_2O_5 \cdot CO_2$ (approx)	1 340.0
Borax	$Na_2B_4O_7 \cdot 10H_2O$	381.4
Boric acid	H_3BO_3	61.8
Boric oxide	B_2O_3	69.6
Calcite (see calcium carbonate)		
Calcium borate (colemanite)	$Ca(BO_2)_2 \cdot 2H_2O$	161.7
Calcium carbonate (whiting)	$CaCO_3$	100.1
Calcium chloride	$CaCl_2 \cdot 6H_2O$	219.1
Calcium chloride (anhydrous)	$CaCl_2$	111.0
Calcium fluoride (fluorspar)	CaF_2	78.1
Calcium hydroxide	$Ca(OH)_2$	74.1
Calcium orthophosphate	$Ca_3(PO_4)_2$	310.3
Calcium oxide (lime)	CaO	56.1
Calcium sulfate (gypsum)	$CaSO_4 \cdot 2H_2O$	172.2
Carbon dioxide	CO_2	44.0
Chromium oxide	Cr_2O_3	152.0
Clay (kaolinite, china clay)	$Al_2Si_2O_3(OH)_4$	258.1
Cobaltic chloride	$CoCl_3$	165.3
Cobalt(II, III) oxide	Co_3O_4	240.8
Cobalt(III) oxide	Co_2O_3	165.9
Cobaltous acetate	$Co(CH_3COO)_2 \cdot 4H_2O$	249.0
Cobaltous carbonate	$CoCO_3$	119.0
Cobaltous chloride	$CoCl_2 \cdot 6H_2O$	238.0
Cobaltous nitrate	$Co(NO_3)_2 \cdot 6H_2O$	291.1
Cobaltous oxide	CoO	74.9
Cobaltous phosphate	$CO_3(PO_4)_2 \cdot 3H_2O$	420.9
Cordierite	$Mg_2Al_4Si_5O_{18}$	584.9
Corundum (see alumina)		
Cryolite	Na_3AlF_6	285.0
Cupric carbonate (basic)	$CuCO_3 \cdot Cu(OH)_2$	221.2
Cupric chloride	$CuCl_2 \cdot 2H_2O$	170.5
Cupric hydroxide	$Cu(OH)_2$	97.6

Table IX. Continued

Mineral name	Mineral formula	Formula weight (lb)
Cupric nitrate	$Cu(NO_3)_2 \cdot 6H_2O$	295.7
Cupric oxide	CuO	79.6
Cupric sulfate	$CuSO_4 \cdot 5H_2O$	249.7
Cuprous chloride	$CuCl$	99.0
Cuprous hydroxide	$Cu(OH)$	80.6
Cuprous oxide	Cu_2O	143.1
Cuprous sulfate	$Cu_2SO_4 \cdot H_2O$	225.2
Diopside	$CaSiO_3 \cdot MgSiO_3$	216.6
Dolomite	$CaCO_3 \cdot MgCO_3$	184.4
Feldspar (see albite, anorthite, orthoclase)		
Ferric chloride	$FeCl_3$	162.2
Ferric hydroxide	$Fe(OH)_3$	106.9
Ferric oxide (hematite)	Fe_2O_3	159.7
Ferric sulfate	$Fe_2(SO_4)_3 \cdot 9H_2O$	562.0
Ferro-terric oxide (magnetite)	Fe_3O_4	231.5
Ferrous carbonate (siderite)	$FeCO_3$	115.8
Ferrous oxide (wustite)	FeO	71.8
Ferrous sulfate	$FeSO_4 \cdot 7H_2O$	278.0
Ferrous sulfide	FeS	87.9
Flint (see silica)		
Gypsum (see calcium sulfate)		
Ilmenite	$FeTiO_3$	151.9
Kaolinite (see clay)		
Kyanite	Al_2SiO_3	162.3
Lead borate	$Pb(BO_2)_2 \cdot H_2O$	310.9
Lead carbonate	$PbCO_3$	267.2
Lead carbonate basic (white lead)	$2PbCO_3 \cdot Pb(OH)_2$	775.6
Lead chloride	$PbCl_2$	278.1
Lead dioxide	PbO_2	239.2
Lead oxide (litharge)	PbO	223.2
Lead oxide (red lead)	Pb_3O_4	685.6
Lithium carbonate	Li_2CO_3	73.9
Magnesium carbonate (magnesite)	$MgCO_3$	84.3
Magnesium chloride	$MgCl_2 \cdot 6H_2O$	203.3
Magnesium oxide (magnesia, periclase)	MgO	40.3
Manganese dioxide	MnO_2	86.9
Manganous carbonate	$MnCO_3$	114.9
Manganous oxide	MnO	70.9
Microcline (see orthoclase)		
Mullite	$Al_6Si_2O_{13}$	425.9
Nickel chloride	$NiCl_2$	129.6
Nickel oxide	NiO	74.7
Niter (saltpeter) (see potassium nitrate)		
Orthoclase (potash spar)	$K_2O \cdot Al_2O_3 \cdot 6SiO_2$	556.8
Potash spar (see orthoclase)		
Potassium carbonate	K_2CO_3	138.0

Table IX. Continued

Mineral name	Mineral formula	Formula weight (lb)
Potassium chloride	KCl	74.6
Potassium chromate	K_2CrO_4	194.2
Potassium dichromate	$K_2Cr_2O_7$	294.2
Potassium ferrocyanide	$K_4Fe(CN)_6 \cdot 3H_2O$	422.4
Potassium hydroxide	KOH	56.1
Potassium nitrate (niter)	KNO_3	101.1
Potassium oxide (potash)	K_2O	94.2
Potassium permanganate	$KMnO_4$	158.0
Pyrophyllite	$Al_2Si_4O_{10}(OH)_2$	360.3
Quartz (see silica)		
Silica (quartz, flint)	SiO_2	60.1
Silicic acid	H_2SiO_3	78.1
Sillimanite	Al_2SiO_5	162.3
Soda ash (see sodium carbonate)		
Soda spar (see albite)		
Sodium bicarbonate	$NaHCO_3$	84.0
Sodium carbonate (anhydrous)	Na_2CO_3	106.0
Sodium carbonate (hydrated) (soda ash)	$Na_2CO_3 \cdot 10H_2O$	286.2
Sodium chloride (salt)	NaCl	58.4
Sodium chromate	$Na_2CrO_4 \cdot 10H_2O$	342.2
Sodium dichromate	$Na_2Cr_2O_7 \cdot 2H_2O$	298.0
Sodium hydroxide (caustic, lye)	NaOH	40.0
Sodium nitrate (soda niter)	$NaNO_3$	85.0
Sodium oxide (soda)	Na_2O	62.0
Sodium silicate	variable $Na_2O:SiO_2$ ratios	
Sodium sulfate (salt cake)	$Na_2SO_4 \cdot 10H_2O$	322.2
Spinel	$MgAl_2O_4$	142.2
Strontium carbonate	$SrCO_3$	147.6
Strontium oxide	SrO	103.6
Sulfur dioxide	SO_2	64.1
Sulfur trioxide	SO_3	80.1
Talc	$Mg_3Si_4O_{10}(OH)_2$	379.3
Tin chloride (stannic)	$SnCl_4$	260.5
Tin chloride (stannous)	$SnCl_2$	189.6
Tin oxide (stannic)	SnO_2	150.7
Tin oxide (stannous)	SnO	134.7
Titania (rutile, anatase)	TiO_2	80.1
Uranium dioxide	UO_2	270.0
Uranium oxide	U_3O_8	842.0
Uranium trioxide	UO_3	286.0
Wollastonite	$CaSiO_3$	116.2
Zinc carbonate	$ZnCO_3$	125.4
Zinc oxide	ZnO	81.4
Zinc sulfate	$ZnSO_4 \cdot 7H_2O$	287.6
Zirconia	ZrO_2	123.0
Zirconium silicate (zircon)	$ZrSiO_4$	183.3

*The glaze binders are starch, dextrin, gum arabic, gum tragacanth, CMC (carboxymethylcellulose) and sugar.

References

[1]"Glazes"; Lead in Ceramics. Lead Industries Association, New York, 1956.

[2]Cullen W. Parmelee; p. 4 in Ceramic Glazes. Cahners Books, Boston, 1973 (revised and enlarged by C.G. Harmon).

[3]Frank Hamer; p. 143 in The Potters Dictionary of Materials and Techniques; Watson-Guptill, New York, 1975.

[4]F. Singer and W.L. German; p. 21 in Ceramic Glazes. Borax Consolidated Limited, London, 1966 (revised 1980 by D.C. Maynard, Borax Holdings Limited, London).

[5]Reference 2, pp. 329–38.

[6]J.T. Jones and M.F. Berard; Ceramics: Industrial Processing and Testing, Chapter 8. Iowa State University Press, Ames, IA, 1978.

[7]Reference 2, pp. 49 and 59.

[8]Reference 3, p. 102.

[9]J.D. Hinton and C.L. Williams, Jr., "Tests for Extractible Lead in Frits and Glazes," Am. Ceram. Soc. Bull., 55, [11] 986–8 (1976).

[10]S.B. Gross, "Oral and Inhalation Lead Exposures in Human Subjects," Lead Industries Association, 1979.

[11]Control of Lead at Work, Health and Safety at Work Act, HMSO, 1978.

[12]"Lenox China, Inc: Application for Variance; Interim Order," Fed. Regist., 47, No. 137 (1982).

[13]F. Singer and S.S. Singer, pp. 553, 554 in Industrial Ceramics. Chapman and Hall, London, 1979.

[14]Annual Book of ASTM Standards, Part 17. ASTM, Philadelphia, PA, 1982.

[15]International Standard, Ceramic Ware in Contact With Food—Release of Lead and Cadmium—Part 1: Method of Test, Part 2: Permissible limits, ISO 6486. International Organization for Standardization, 1981.

[16]Reference 1, pp. 5 and 6.

[17]Reference 1, pp. 7–10.

[18]"Sampling and Testing Structural Clay Brick and Tile," Ref. 14, Part 16, C-67–81.

[19]"Crazing Resistance by Thermal Shock Method," Ref. 14, C554.

[20]"Crazing Resistance by Autoclave Treatment," Ref. 14, C424.

Hobby Glazes

Hobby ceramics is a relatively new area within the whitewares field. Body compositions, kilns, and firing ranges have evolved that have brought a measure of standardization to the industry.

Most hobby ceramic pieces are made with earthenware body of high talc content. The percentage of talc may vary from 50 to 64%, and the shapes are formed by slip casting.

The two fire process is most common with a cone 05-04 bisque and a cone 06-05 oxidizing glaze fire (see Table I). Firing is accomplished in periodic electric kilns, varying in size from 0.03 to 0.34 m^3 (1–12 ft^3). The firing time to the maximum temperature varies from 3 to 6 h.

Hobbyists create individual pieces and glaze manufacturers produce a wide variety of products for their use. Since the amount of any one glaze used on a piece is small, the glaze is applied by brush. Brushing properties are achieved by the addition of a cellulose gum to the glaze batch. The glazes are milled in water and bottled in a ready-to-use form.

The desired fired properties of a hobby glaze are the same as those sought when formulating any glaze — craze resistance, wide firing range, color stability, etc. The problems are more acute because the hobby glaze will be applied to many different shapes and fired in a variety of kilns with ever-changing kiln loadings.

The manufacturer seeks to produce the most foolproof glaze, often foregoing economies that can be practiced in potteries where the glaze is used under controlled conditions.

The major portion of the glaze is fritted and the most successful are those bearing lead (Table II). No combinations of elements equal the fluxing proper-

Table I. Typical Cone 06-05 Hobby Glazes*

	Rutile float	Clear	Transparent matte	White lava
Frit A	75.0			
Frit B		90.0		
Frit C			65.0	
Frit D				87.5
Rutile	15.0			
Silica	5.0		6.0	
Zinc oxide	5.0			
Bentonite	1.0			1.0
Kaolin		10.0	8.0	
Wollastonite			20.0	
Tricalcium phosphate				12.5

*Courtesy of Mobay Chemical Corp., Baltimore, MD.

Table II. Oxide Percentage Compositions*

	K_2O	Na_2O	CaO	MgO	BaO	ZnO	PbO	Al_2O_3	B_2O_3	P_2O_5	SiO_2	ZrO_2	TiO_2
Rutile float	1.1					5.0	53.1	2.1			24.3		14.4
Clear		3.4	3.9				27.9	7.3	11.6		44.7	1.1	
Transparent matte	1.5	0.3	13.7	0.4	10.0		20.8	5.8	8.2		39.4		
White lava	1.0	2.4	12.7				3.0	5.0	21.0	5.7	49.1		
Frit A		1.5					71.2	2.4			24.9		
Frit B		3.6	4.5				31.0	3.4	13.0		43.5	1.0	
Frit C	2.2	0.6	6.5			15.0	30.9	3.7	12.3		28.8		
Frit D	1.1	2.8	6.9				3.5	5.6	24.2		55.9		

*Courtesy of Mobay Chemical Corp., Baltimore, MD.

ties of lead in a low-fire glaze, while maintaining low expansion, low surface tension, and low viscosity over a wide firing range. Much development work has been done in an attempt to replace lead, but the results have had little success.

The potential hazards of lead must be carefully considered. For protection of the hobbyist, lead must be used in a fritted form, and only frits with the lowest practical acid solubility should be used. Air-born dust is of minimal concern because the glazes are supplied in a water suspension. Upon drying, the cellulose gum serves as a binder and produces a relatively dust-free unfired glaze film.

Lead-bearing glazes that pass the current regulations for use on food containers can be made in the cone 06-05 firing range. Many hobby glazes are for decorative use only. It is the practice of the hobby glaze manufacturers to label their products by stating whether they are or are not satisfactory for food containers.

5

Enamels

The primary use of lead compounds in porcelain enamel compositions is as a flux. As fluxing compounds, they offer the advantage of being able to be used in relatively large quantities without causing a dull finish by the formation of crystalline compounds. In fact, the use of lead compounds generally leads to an increase in brilliance and smoothness of the fired enamel. In addition, their use normally leads to increased resistance to chipping, improved elasticity, and increased corrosion resistance. Consequently, lead compounds have found an ongoing important place in the porcelain enameling industry, particularly since energy considerations have made lower firing temperatures more desirable.

Cast Iron

In dry-process enameling on cast iron, lead compounds are commonly used both in the ground coat and the cover coat enamels. With each type of enamel, the high fluxing properties of lead are a definite advantage. This is particularly evident in the resmelting of the cover coat powder reclaimed from the dusting operations. With the antimony-opacified type of cover coat enamel, lead oxide in excess of 10% tends to produce yellow enamels. Typical ground coat compositions are shown in Table I and cover coat compositions in Table II. Lead compounds are also often used in wet-process enameling of cast iron.

Steel

In developing compositions for the enameling of sheet steel, the use of lead compounds has been primarily for special enamels. Typical of these special

Table I. Dry Process Ground Coat Enamels for Cast Iron*

	Batch weights (%)				
	Frit A	Frit B	Mill additions	Parts	Parts
Lead bisilicate	3.0		Frit A and B	100	100
Litharge		4.0	Clay	5	15
Dehydrated borax	18.6	21.0	Iron oxide	1	
Feldspar	65.1	45.0	Flint		20
Quartz	12.7	30.0	Zircon	0.5	
Manganese dioxide	0.6		Soda ash		10
			Borax	2	

*Firing temperature range for frits A and B is 900–930°C (1650–1706°F).

Table II. Dry-Process Cover Coat Enamels for Cast Iron*

| | Batch weights (%) | | | |
| | Regular enamels | | Acid-resisting enamels | |
	Frit A	Frit B	Frit C	Frit D
Litharge	19.0		12.5	6.4
Lead bisilicate		7.2		
Quartz	34.3	1.3	36.0	33.7
Soda ash	17.6		14.9	13.0
Sodium nitrate	5.0	3.3	7.4	4.9
Fluorspar	2.0	7.0		2.0
Sodium antimonate	11.0	10.0	8.7	9.6
Calcium carbonate	2.8	1.2	4.1	4.4
Sodium silicofluoride	1.0		2.0	
Bone ash	1.0			
Cryolite		1.2		2.0
Feldspar		27.7		
Zinc oxide		9.3		00
Barium carbonate		5.0		
Titanium dioxide			6.0	5.4
Hydrated borax	6.3	26.8	8.4	18.6

*Firing temperature range for frits A–D is 840–870°C (1544–1598°F).

enamels are blacks both regular and acid resisting. Compositions of this type are shown in Table III.

In recent years significant increases in the cost of energy have renewed interest in the development of porcelain enamels for steel that matures below 1400°F (760°C). The reduction of firing temperatures offers the obvious reduction in energy cost, but in addition, it provides the opportunity to broaden the types of low-cost cold-rolled steel that can be successfully enameled. Design considerations as related to strength and warpage for enamels that are fired in the 815–870°C (1500–1600°F) range can oftentimes be changed to lighter gages when porcelain enamels are matured below the transformation temperature (strain point) of the steel. Lead compounds are used as a flux addition in such compositions.

Lead-containing enamels of the same type as those being used for aluminum (see aluminum) are also employed for aluminum-clad steel and for decorative purposes on stainless steel.

Developments in recent years have provided industry with suitable lead-bearing frits for porcelain-enameled aluminum, allowing for its use in widely varying applications. The essential components of many of these enamels are lead oxide, silica, and the alkali group (K_2O, Na_2O, and Li_2O), particularly lithium oxide (Table V). Within this system, an increase of the ratio of lead to silica and alkali will increase the gloss and flow of the enamel but with some loss of acid resistance. The alkali group accounts for the needed high coefficient of

Table III. Typical Steel Enamel Compositions Containing Lead Compounds*

	Batch weights (%)		
	Regular black		Acid-resisting black
	Frit A	Frit B	Frit C
Red lead	2.7	6.5	5.0
Hydrated borax	29.3	30.9	21.8
Soda ash	6.2	1.9	14.9
Sodium nitrate	4.9	4.4	4.0
Quartz	19.4	12.5	37.4
Barium carbonate			3.0
Feldspar	26.6	24.2	
Cryolite	6.9		2.5
Titanium dioxide			7.9
Fluorspar		5.4	
Manganese dioxide	3.1	2.9	3.0
Cobalt oxide	0.9	10.2	0.5
Black oxide		1.1	
Mill additions	**Parts**	**Parts**	**Parts**
Frit A, B, or C	100	100	100
Clear clay	4.5	4.5	4.5
Black oxide	2.3	2.3	2.3
Water	45	45	45
Potassium carbonate			0.25–0.5

*Milled to 2–4% residue on 100 mesh. Firing temperature for Frits A–C is 816–838°C (1500–1540°F) for 3–4 min.

expansion. As in all ceramic systems though, it is the "proper" balance of lead, silica, and alkali which achieves low firing temperature, high gloss, and high thermal expansion with excellent chemical stability.

Usually, some of the silica will be replaced by titania. Smelted in the frit, titania does not increase acid resistance as might be expected, but is added to increase aging stability in water suspensions. Titania mill additions, however, will increase acid resistance. Some other modifications of the basic frit that have become necessary are as follows:

(1) Antimony pentoxide is added in small percentages to reduce the tendency of the enamels to darken where there are traces of organic matter, not, as is the usual practice, for opacification.
(2) Boric oxide is added in small percentages to reduce tearing and crawling.
(3) In practice, opacity is obtained by titania mill additions. Titania in the frit will produce some opacity, but high opacity cannot be obtained in this way.

Table IV. Typical Mill Formula

Component	Composition (wt%)
Frit (also flake or powder)	100
Boric acid	3
Potassium hydroxide	2.6
Potassium silicate	6.2
Titanium dioxide	8.0
Color pigment (oxide) A	2.5
Color pigment (oxide) B	1.1
Water	34

Table V. Some Representative Frit Compositions for Aluminum

Frit	Composition (wt%)				
PbO	38.1	34.7	35.4	36.4	42.5
SiO_2	25.3	31.5	23.4	34.9	28.2
Li_2O	2.2	2.6	2.0	2.2	2.4
Na_2O	10.2	11.1	9.5	13.4	9.8
K_2O	8.3	8.1	7.7	2.2	7.4
TiO_2	8.7	10.7	8.1	9.3	9.7
Sb_2O_3	2.9		1.8	1.6	
B_2O_3	4.3		4.0		
ZrO_2		1.3			
CdO			8.1		

(4) These enamels are stable with all coloring oxides except the cadmium sulfoselenide red and orange pigments. However, 8% cadmium oxide (CdO) in the frit will stabilize these pigments.

Base Metals

A fairly limited range of aluminum alloys can be porcelain enameled, and it is recommended that users discuss selection of base metals with an aluminum supplier or reputable enameler before specifying the base metal. Aluminum companies produce specially developed alloys for porcelain enameling. They are formulated for various end uses and to yield various properties after being enameled. In most cases, the companies that developed them have labeled these enameling alloys with their own designations. There are several readily available alloys carrying industry-wide designations that are recommended for enameling. They are listed in the following table:

	sheet	extrusion	casting
	1100	6061	43
	3003	7104	
	6061		356

Of the wrought alloys, only 6061 alloy is heat treatable. Because of its higher strength, 6061 alloy has better handling characteristics before and during porcelain enameling. It is stronger after porcelain enameling. The non-heat-treatable alloys are easier to form before porcelain enameling and are used for small parts for which the amount of distortion and low strength encountered after firing are acceptable; however, non-heat-treatable alloys are unsuitable for more than one coat of porcelain because of crazing after a second firing.

Metal Preparation

The preparation of aluminum base metal for porcelain enameling involves the removal of soil which can be done by alkaline cleaning or vapor degreasing. Parts made of non-heat-treatable aluminum alloys require only the removal of soil. Parts made of heat-treatable aluminum alloys involve additionally the removal of surface oxide and the application of a chromate coating. The final dried product should be free of surface moisture. The surface to be coated must be free of all foreign material to allow wetting with the enamel slip, as sprayed, and to avoid defects due to reaction during firing.

Enamel Slip Preparation

Aluminum enamels are prepared for application by grinding the frit and mill additions in porcelain-lined ball mills (Table IV). These slips are ground considerably finer than typical steel enamel slips. Fine grinding is necessary because of the need to melt the frit particles at a low maturing temperature. A typical fineness would be less than 0.5% coarser than 44 μm.

Applying the Enamel

The properly prepared metal and milled enamel are brought to the spray area for application of the coating. Various spray setups are used, depending on the type of product and volume. These vary from hand spraying in individual booths and loading onto a furnace conveyor or charging mechanism to loading on a conveyor that feeds through the spray area onto a furnace conveyor.

All spraying is with air atomizing guns, performed by hand or mechanized. For example, spindle machines may be used for parts such as pots and lids and reciprocating units for basically flat panels. Aluminum porcelain enamels fire to essentially the same surface as deposited in the wet spray, i.e., they do not flow out during firing. Therefore, the spray must be adjusted to produce the desired surface characteristics.

When screened decoration is required, special screening pastes are applied over a fired coating. Screen-decorated ware is air-dried to remove volatile components before firing the work.

The sprayed coating cannot be completely dried before firing. However, a short exposure to the air dissipates surface water film. This allows a reasonable time to transfer ware to the furnace for firing.

Firing the Finish

The coating is fused into a glassy coating at 525–550°C (980–1020°F). The rate of heating and uniformity of temperature from area to area are important. Firing is performed in a normal atmosphere.

The type of furnace heat is a matter of economics and ability to control temperature. Both gas and electric furnaces are used. Porcelain-enameled aluminum can be fired in either convection or radiation units. Convection generally is used to obtain temperature control and uniformity of temperature in the work load.

Various systems may be used for firing. These include continuous, intermittent, and box furnaces. Work may be handled by fork, monorail, roller hearth, or link belt. Most work is run on continuous convection furnaces with link belt conveyors. The time cycle for firing depends on the thickest section of the work being fired, with heavier sections requiring longer firing times.

For general purposes, minimum time in the heated zone of a furnace is 8 min. This gives 4 min to heat and a 4-min soak at temperature. The 8-min period is common for 0.054-in. (0.137-cm) sheet. Heavy sections such as castings may require as much as 15 min in the heated section of the furnace.

As production is brought from the furnace (first time for single coat or second time for two coat or screen decorated), it is ready for inspection and quality verification.

An in-depth description of all aspects of porcelain enameling of aluminum can be found in a series of Porcelain Enamel Institute (PEI) bulletins on the subject.[1-6]

Properties of Enameled Aluminum

Adherence

Resistance of spalling, a defect characterized by separation of porcelain enamel from the base metal without apparent external cause, is the indicator used to measure adherence of porcelain enamel on aluminum. Spalling can result from the use of improper alloys or enamel formulations, incorrect pretreatment of the base metal, or faulty application and firing procedures.

ASTM C703, "Spalling Resistance of Porcelain Enameled Aluminum," outlines two methods for determining resistance to spalling. Method A, using a 5% solution of ammonium chloride, requires a 96-h immersion of the test specimen. Method B uses a 1% solution of antimony trichloride and requires a 20-h immersion of the test specimen.

The spall test is a pass/fail test, with failure determined by (a) the existence of spall areas of specified dimensions at specimen edges or (b) the existence on the specimen interior of spots exceeding specified dimensions or a spot level exceeding a specified density, usually spots/m^2 (spots/ft^2).

Surface Texture and Color

Almost any degree of surface reflectivity, from a very dull matte to extremely high gloss, can be obtained. Possible surface variations are innumerable. There is no limitation to the range of light-fast colors. The durable finish minimizes maintenance requirements.

Reflectance

Reflectance for white light ranges up to 88%, with total reflectance as high as 90%.

Abrasion Resistance

Abrasion resistance is one of porcelain-enameled aluminum's notable features. Porcelain-enamel coating for aluminum offers about 10 times the abrasion resistance afforded by organic finishes such as alkylamine paints as measured by standard test methods.

Chemical Resistance

Chemical resistance of porcelain-enameled aluminum varies with enamel formulations and firing conditions, but porcelain enamels that meet PEI specifications have high chemical durability. They are colorfast and show little change of gloss after years of exposure to weather. Tests conducted by the National Bureau of Standards[7] and the Porcelain Enamel Institute have shown a high correlation between a porcelain enamel's acid resistance rating and its actual weather resistance.[8] Formulations with boiling citric acid weight loss figures under 20 mg in.$^{-2}$ have excellent weather resistance.

Resistance to Impact

Resistance to impact is considered to be one of porcelain-enameled aluminum's invaluable properties. Samples have averaged five impacts in standard tests before loss of coatings. Even then, no sharp edges, progressive spalling, or flaking was noted. Because the base metal does not corrode, small damaged spots do not require repair.

Torsion Resistance

A torsion resistance rating of 200+°C for 3-mil (7.6 × 10^{-2} mm) thickness of porcelain enamel is given according to the test described by the Porcelain Enamel Institute, which illustrates the formability of porcelain-enameled aluminum. This test employs right-angle-shaped specimens, 12 in. (30.5 cm) long and with a ³⁄₁₆-in. (0.5-cm) radius at the right-angle bend. The specimen is twisted until coating failure occurs at the radius of the bend.

Thermal Shock

Panels heated to 538°C (1000°F) have been plunged into cold water with no cracking or flaking of the enamel coating.

Salt Water Corrosion Resistance

Enameled surfaces have shown no deterioration after 480 h in a 20% salt spray test.

Dielectric Strength:

The enamel has a voltage breakdown rating of 500 V mil^{-1} (196.8 × 10^3 V/cm), making it suitable for insulating and shielding purposes.

Uses of Porcelain-Enameled Aluminum

Architectural

The porcelain-enameled aluminum industry was initially launched by architectural applications. Although porcelain-enameled aluminum's properties make it a frequent vehicle for bringing permanent color, interesting design, and freedom from maintenance to copings, fasciae, canopies, and trim, its widest architectural use is in panels. These may be unsupported sheets, veneer, or composition core sandwich panels for curtain walls, window walls, or spandrel-type walls. They are supported by secondary framework, which may be sash frames or a mullion bar system attached to the main structural frame.

Sandwich panels are core materials laminated between two facings, one or both of which are porcelain enameled. Core material may be composed of a variety of readily available substances. For large spandrels, thin-gage porcelain-enameled aluminum can be bonded to heavier aluminum sheets to achieve flatness.

Adhesive bonding distributes stress loads more evenly over entire joined areas than do mechanical fasteners which may concentrate stress locally. If the adhesive is not sufficiently elastic, dissimilar materials having different coefficients of expansion may create stresses great enough to warp the assembly or destroy the bond with a brittle adhesive. Several useful general purpose industrial adhesives are available. Neoprene contact types offer a high-strength, resilient, heat-and-water-resistant bond.

When mechanical fasteners are used, care should be taken to avoid contact between dissimilar metals because of galvanic action. They should be separated by insulators or be treated with protective coatings. Because porcelain enamel is a permanent finish, fasteners should also be corrosion or stain resistant.

The excellent field-handling characteristics of porcelain-enameled aluminum are attributable to one of its chief properties, a high degree of formability. There is no progressive spalling of the tightly adherent coating from a cut edge or damaged spot. This tightly adhering, flexible coating permits porcelain-enameled aluminum to be cut, drilled, sheared, and even, to some extent, formed after enameling. The material can be cut and fitted and holes can be drilled for fasteners or wiring at the construction site. Coating of sawed edges and repair of small damaged spots is often unnecessary, thanks to the base metal's resistance to corrosion. No special skills are needed to drill, saw, or cut porcelain-enameled aluminum, but these operations should be carried out with care. The panels should be backed up firmly, masking tape should be used over the working area, saws should operate at high speed, and punching that requires strong holding devices should be avoided because of possible damage to the porcelain enamel surrounding the holes.

Porcelain-enameled aluminum can be formed in the field to an appreciable extent. Welding can be performed in the field on the reverse sides of porcelain-enameled aluminum if extreme care is taken not to burn through the lighter gages. Electric welding should be used and heat should be held to a minimum, well below 1000°F. Because even the most careful welding may result in some distortion (although often barely noticeable) welds should be planned wherever possible as part of the design and be carried out before enameling. Porcelain-enameled aluminum is characterized by torsion resistance and high flexural strength that permit the material to be cut, drilled, and formed in the field.

Color and Texture

Porcelain-enameled aluminum is available to the architect and sign designer in an almost infinite variety of colors and finishes. Multicolor effects can be achieved by using several colors on a single panel. The degree of gloss ranges from a mirror-like finish to a matte finish. Highlights may be further diminished by use of an embossed aluminum sheet.

Incorporation of fade-proof colors into structures is one of the uses for porcelain-enameled aluminum best known among architects. Many standard colors are available and porcelain-enameled aluminum components can be produced in a wide range of "nonstandard" colors and gloss variations to carry out company images, match insignias, or meet any other architectural or sign need.

Design and Shapes

Almost any shape that can be produced in aluminum by drawing, rolling, brake forming, extruding, or casting can be porcelain enameled. Unusual, intricate forms lend distinction to fasciae, louvres, column covers, grilles, and emblems. They can be produced in a variety of gages, and components can be as large as 5 ft (1.5 m) wide and 25–30 ft (7.6–9.1 m) long or longer. Narrow extrusions, often used for trim, are readily porcelain enameled.

Design of aluminum components to be porcelain-enameled generally can be carried out as it would be for nonenameled components. The chief design consideration for enameling is that all angles should be rounded slightly. A minimum radius of ⅛ in. (0.32 cm) should be observed, although an outside radius as small as ¹⁄₁₆ in. (0.16 cm) is sometimes used in consultation with the enameler.

Cookware

The beauty and durability of porcelain enamel has become a part of the aluminum cookware industry.

Cookware fabricated from high-strength aluminum alloy sheet that are coated with porcelain enamel on the outside and nonscratch, antistick coatings on the inside have been well received in the market place. The basic properties of porcelain-enameled aluminum, permanent color, ease of cleaning, heat resistance, corrosion resistance, and abrasion resistance all lend themselves to cookware applications. A wide range of decorative and durable colors are available in porcelain-enamel coatings.

Aluminum cookware made by the casting process with an exterior porcelain enamel finish has also been successful with a history dating back to 1951.

"Recommended Specifications for Porcelain Enamel Finishes on Aluminum Cookware" was published by the Porcelain Enamel Institute, Inc., with specific recommendations for finish, surface durability, coating thickness, and spall resistance.

Miscellaneous

Being highly resistant to salt spray and salt-water corrosion, aluminum porcelain enamels have a market on the seas. Thousands of square feet have been installed as partitions in wet spaces. On the other side of the tonnage displacement scale, many small pleasure craft owners have installed porcelain-enameled hardware on their boats.

Another use of aluminum porcelain enamels is the manufacture of chalk board and marker board. These provide the necessary durability and erasability that is needed for educational and industrial applications.

Highway, informational, and commercial signs are another market for porcelain-enameled aluminum.

References

[1] "Alloy, Design and Fabrication Considerations For Porcelain Enameling Aluminum," Bulletin P-402, Porcelain Enamel Institute, Inc., Washington, DC.

[2] "Pretreatment of Alloys For Porcelain Enameling Aluminum," Bulletin P-403, Porcelain Enamel Institute, Inc., Washington, DC.

[3] "Enamel Preparation, Application and Firing For Porcelain Enameling Aluminum," Bulletin P-404, Porcelain Enamel Institute, Inc., Washington, DC.

[4] "Quality Control Procedures For Porcelain Enameling Aluminum," Bulletin P-405, Porcelain Enamel Institute, Inc., Washington, DC.

[5] "Specifications For Architectrual Porcelain Enamel on Aluminum For Exterior Use," PEI:ALS-105, Porcelain Enamel Institute, Inc., Washington, DC.

[6] "Recommended Specification For Porcelain Enamel Finishes on Aluminum Cookware," PEI:ALS-106, Porcelain Enamel Institute, Inc., Washington, DC.

[7] Margaret A. Baker, "1964 Exposure Test of Porcelain Enamels on Aluminum — Five Year Inspection," NBS Report 10875, National Bureau of Standards, Washington, DB.

[8] "Weathering of Porcelain Enamels on Aluminum, 12-Year Inspection, Exposure Period 1964-1976," Department of Ceramic Engineering, University of Illinois at Urbana-Champaign, 1977.

6
Glass Enamels and Colors

Glass color enamels have been in existence for nearly as long as glass itself. They have been applied to a wide range of products including beverage bottles, pharmaceutical and cosmetic packages, laboratory apparatus, architectural and automotive glass, lighting fixtures, light bulbs, cookware, appliance panels, tumblers, food storage, and tableware.

The means for applying glass enamels varies widely but usually requires the use of a vehicle to temporarily bind glass and pigment particles to the substrate being decorated until subsequent firing fuses them to that surface. These methods include hand painting, spraying, decalcomania, beading, banding or lining, dipping, pad transfer, stamping, gravure, and screen printing.

Firing these enamels can be done in periodic kilns or furnaces, both electric and gas fired; however, more large-scale production firing is accomplished by passing freshly decorated ware through a lehr that preheats the ware, volatilizing organic vehicle constituents before the enamel softens, and then matures the enamel with controlled heating — above the softening point (time and temperature) — before cooling in a controlled manner to anneal the ware and prevent buildup of undue stress.

The softening point, annealing point, and strain point of most glasses on which glass enamels are used are low enough to require a low-melting flux base for the colors being applied. In most cases, this flux is a lead borosilicate glass sometimes referred to as a "frit." Acid, alkali, and sulfide resistance can be increased by the addition of elements such as titanium, zirconium, and aluminum. As the concentration of these elements increases, so does the firing temperature, making it necessary to control the proportions very carefully while developing maximum chemical resistance, a sufficiently low maturing temperature, and maintainance of the proper coefficient of thermal expansion to match that of the glass for which the color is intended.

Pigments, in varying concentrations, are added to the flux base and micropulverized or ball-milled together until the average particle size of the mixture reaches approximately 2–5 μm. Table I lists the colors and the oxides normally used in their preparation.

Glass Enamel Flux Properties

In the development of glass enamels, there are four properties of primary concern: stability of composition, chemical durability, coefficient of expansion, and melting temperature. A brief outline of these properties follows.

Table I. Oxides Used for Glass Enamel Colors*

Color	Oxide
White	Titanium dioxide
Brown	Chromium oxide + zinc oxide + iron oxide
Blues	Cobalt oxide + aluminum oxide
Blacks	Cobalt oxide + iron oxide + chrome oxide
Yellows	Lead oxide + antimony oxide or cadmium process[†]
Greens	Cobalt oxide + chromium oxide
Reds	Cadmium process[†]
Orange	Cadmium process[†]

*When enamels with these pigments are used in conjunction with lead-based enamels containing no cadmium, the reaction forms black lead sulfide discoloration. Therefore, it is a common practice to incorporate some cadmium oxide in all colors to stabilize them against the formation of lead sulfide or lead selenide compounds.

[†]Cadmium process refers to the reaction of cadmium metal with selenium metal and/or sodium sulfide to form cadmium sulfoselenide pigments which range from the bright yellow cadmium sulfide to brilliant oranges and reds.

Stability of Composition

The composition of the flux should be such that it will not crystallize under normal firing conditions, and its chemical reactivity should preclude formation of compounds that would adversely affect the color.

Chemical Durability

There is an increasing demand for enamels that have extremely high resistance to acids and alkalies.

Enamels that come in contact with food must resist various acids and sulfides and concurrently resist strong alkaline detergents used in automatic dishwashers. With the preponderance of these machines in use in most households, the application of tumbler enamels, commonly referred to as "soft AR" (acid resistant) colors, has decreased significantly. Multitrip beverage bottle enamels have also been faced with strong, hot alkaline solutions as well as hot acids which remove iron stains resulting from metal closures. High acid-resistant colors serve best in the fields of structural and automotive glass where they are subjected to attack by weather.

Coefficient of Thermal Expansion

An enamel that does not "fit" the expansion of the host substrate can cause crazing due to stresses in the enamel or even glass fracture if the stresses are too great and penetrate far enough into the substrate. As a general practice, it can be stated that an enamel, to have a good fit on the glass to which it is applied, should have a linear coefficient of expansion about three points lower than the expansion of the substrate glass. For example, if a glass has a coefficient of $81 \times 10^{-7}/°C$, then the enamel should have a coefficient of $78 \times 10^{-7}/°C$ or less. There is little chance of having too low a coefficient of the enamel which could cause crazing or weakening of the glass structure.

As the coefficient of expansion of the enamel is increased, the strength falls off rapidly. There is a noticeable weakening of the glass even when the coefficients of glass and enamel are similar. When the expansion of the enamel is 10 points higher than that of the glass, spontaneous crazing will occur and, frequently, breakage of the glass. Thicker enamel applications compound this problem.

Melting Temperature

Since all enamels are applied to articles of preformed glass, it is essential that the enamel have a melting temperature far enough below the softening point of the glass so that the enamel can be fired to a good gloss without causing deformation of the glass article.

It is mandatory that all the above properties be considered whenever decorating enamels are to be applied to a glass article.

Typical Compositions of Enamels

Table II gives the compositions of 10 typical compositions ranging from the very lowest melting, nonresistant type to the higher melting varieties with excellent chemical durability. These compositions will serve to discuss more fully the consideration involved in their design.

Table II. Typical Compositions of Glass Fluxes

Oxide	Flux composition (%)									
	1	2	3	4	5	6	7	8	9	10
PbO	81.8	78.8	66.6	63.2	59.5	55.5	45.5	50.9	56.9	56.03
B_2O_3	10.6	9.6	18.6	8.2	3.0	3.0	2.2	8.8	4.7	2.96
SiO_2	7.6	11.6	14.8	21.8	32.0	29.6	36.3	29.9	25.8	33.36
Al_2O_3				1.5				5.7	0.3	0.06
CdO				4.85	4.0	3.5	4.3	2.9	4.6	3.24
Na_2O				0.3	1.5	4.4	4.8		2.2	2.10
TiO_2				0.05		4.0	4.9	0.4	5.4	2.19
Li_2O							2.0			
ZrO_2				0.03				1.35		
Fe_2O_3				0.04				0.05	0.03	
K_2O				0.03					0.02	
ZnO										0.06
CaO									0.05	
Firing temp (°F)	850	950	1050	1080	1120	1120	1100	1140	1120	1130
Firing temp (°C)	454	510	567	582	604	604	593	616	604	610
Coef. of Exp. $(\times 10^{-7})$ $(cm \cdot cm^{-1}/°C)$	104	85	76	74	78	89	97	54	85	84
Alkali Resistance	Ext. poor	Ext. poor	Ext. poor	Fair	Good	Good	Good	Good	Exc.	Exc.
Lead release (g/cm^2) in 24 h in 4% HAC	Dissolves	Dissolves	Dissolves	6400	667	53	2.7	150	3.75	2.2

Flux No. 1

Compositions such as this, which are extremely low melting due to their high lead content, tend to crystallize if subjected to a high temperature for a prolonged period. When a flux crystallizes, gloss disappears and a large increase in expansion takes place. One such flux increased from an expansion of 75×10^{-7} to 140×10^{-7} after crystallization. When this occurs in this type of composition, it can be corrected by reducing the lead or increasing the silica content. Attempts to make lower melting fluxes than this results in compositions which crystallize during normal firing cycles. Flux No. 1 will dissolve completely in 5% acetic acid in only 10 min. Alkali resistance is also extremely poor.

Flux No. 2

The reduction in lead and increase in silica content increases the firing temperature and decreases the coefficient of expansion. However, there is only slight improvement in chemical durability. Fluxes of this type will find some application where low firing is essential and chemical durability is not an important factor.

Flux No. 3

The lead is further reduced and the silica and boric oxide contents increased, resulting in increased melting temperature and lower expansion. However, this composition still has very poor chemical durability.

Flux No. 4

This was an early attempt at improving chemical durability. This composition is still quite soluble in dilute acids and is subject to attack by various alkaline detergents. Perhaps the outstanding characteristic is its low coefficient of expansion.

Cadmium oxide, which is a good fluxing agent and behaves similarly to lead oxide, is used in lead-containing enamels for the purpose of stabilizing cadmium sulfoselenide red pigments. When this pigment is put in a lead flux containing no cadmium oxide, the mixture will fire out gray to black.

Flux No. 5

This composition is an excellent example of the high-fluxing power of sodium oxide. In comparison with No. 4, the melting temperature was increased only 22°C in spite of the substantial reduction in boric oxide and increase in silica contents. Fluxes of this type have fairly good alkali resistance.

Flux No. 6

Continued development found that introduction of titania into lead fluxes greatly increased the acid resistance. Fluxes No. 5 and 6 are quite similar except that 4% titanium dioxide has been added, and the alkali has been increased in order to accommodate this addition of a more refractory material with no increase in melting temperature. Lead solubility was decreased significantly.

Flux No. 7

This composition employed lithia in combination with sodium oxide as secondary fluxes. Because there are so few glasses having a coefficient of expansion this high, the No. 7 flux would have a limited field of application. However, it exhibits very low lead solubility and possesses excellent weathering characteristics, as is required in the field of structural glass.

Flux No. 8

This composition is typical of a low expansion type used for enamels applied to borosilicate ware, especially ampules, vials, and serum bottles. It has a higher firing temperature due to addition of alumina, which is a more refractory material, and the absence of alkali oxides. It is necessary to use alkali-free fluxes on borosilicate glass, because the alkali metal ions penetrate into borosilicate surfaces, thereby inducing stress and weakening the glass. Chemical durability is good for both acidic and alkaline conditions.

Flux No. 9

Increased levels of titanium dioxide and boric oxide, together with a reduction of sodium oxide, produced a substantial change in chemical durability. This flux is near the state of the art in providing excellent acid resistance and much improved alkali resistance compared to flux No. 7. This flux will find use for tumbler enamels, which must withstand highly alkaline dishwasher detergent attack.

Flux No. 10

The coefficient of expansion remains stable even though silica content has been increased significantly from that of No. 9. This was balanced by a reduction in titania. This is an excellent flux base but fires a little too high and could cause deformation of heavy-bottom glasses and slumping of stemware.

Summary

Lead oxide is the primary flux in glass enamels. Boric oxide is only moderately active as a fluxing agent; however, it is extremely valuable for the reduction of thermal expansion. It lowers chemical durability very rapidly as it is increased.

Cadmium oxide is valuable as a fluxing agent and essential as a stabilizer for the production of bright reds and cadmium yellows.

Both sodium and lithium oxides are valuable as fluxing agents, providing for the addition of larger amounts of silica and the introduction of titania and zirconia. The alkali oxides, however, raise the coefficient of expansion very rapidly and reduce chemical durability.

Silica contributes materially to chemical durability, especially acid resistance. It lowers the coefficient of expansion and increases melting temperature.

Titania is valuable for the improvement of acid resistance.

Zirconia is essential to the production of enamels having good alkali resistance.

Lead-Free Enamels

Lead-free glass enamels were introduced a few years ago and have found application on lighting fixtures and liquor and cosmetic containers. Because their chemical resistance is much lower than that of standard lead-based enamels, they are finding very limited use. No lead-free systems are currently known to the industry that have all the requirements necessary to allow for use on cookware, tableware, lab ware, and other areas that demand excellent chemical resistance.

Advantages claimed for lead-free enamels are that they are temperature stable, their colors are brighter, they are more opaque (cannot be made translucent), and they need not be cadmium stabilized, since there is no lead to form lead sulfide when cadmium sulfide pigments are used.

Papers presented the past 2 years at the Society of Glass Decorators by Committee B-10 offers the consensus that lead-free systems are as good as suppliers can make right now. A marvelous discovery will be needed to increase durability and broaden the application areas. The chances of this happening was felt to be quite remote.

Table III compares typical makeup of the two types of enamels.

The reason for the advent of lead-free enamels stems from environmental concerns over lead in the atmosphere during manufacture and during spraying of lead-based enamels. An incident involving lead- and cadmium-containing enamels on promotional glassware also contributed to manufacturers taking a hard look at the durability of standard enamels.

Since July 1977, when the above-mentioned incident occurred, a concerted effort has been made to improve durability and reduce lead and cadmium release under acidic and alkaline test conditions. New low-lead-release enamels offer resistance to acetic acid leaching at least 20 times better than the soft AR (acid resistant) glass colors in general used only a few years ago. Compositions 9 and 10 in Table II fit this category.

A voluntary standard, mutually agreed to by industry and the federal government, established the limits of lead and cadmium release from decorated glass tumblers as 50 ppm Pb and 3.5 ppm Cd.

These amounts of heavy metal release are allowed when the top 2 cm, or the lip and rim area, is leached in 4% acetic acid for 24 h. The resulting metal release is to be determined by atomic absorption analysis according to ASTM Standard Test Method C 927-80 titled "Lead and Cadmium Release Extracted from the Lip and Rim Area of Glass Tumblers Externally Decorated with Ceramic Glass Enamels."

Table III. Comparison of Two Types of Enamels

Lead-free enamel		Lead-containing enamel	
%	component(s)	%	component
10%	Na_2O, K_2O	7.0%	Li_2O
		2.5%	Na_2O
4%	Al_2O_3	0.5%	Al_2O_3
22%	SiO_2	28.0%	SiO_2
24%	B_2O_3	6.0%	B_2O_3
40%	ZnO, Li_2O, CaO, BaO, SbO	56.0%	PbO

Historical background, including the Federal Register announcement of this standard, the governmental task force report, and a listing of qualified testing laboratories, is covered in a Society of Glass Decorators publication titled "Glass Tumbler Decorations, Suggested Guidelines for Complying with the Voluntary Product Standard involving Lead and Cadmium in Glass Tumbler Decorations."

Methods of Application

Direct Methods

Ceramic glass enamels can be applied by many standard decorating or printing techniques. By far, the most universal would be *screen printing*. Automatic multicolor machines are capable of speeds in excess of 100 bottles, jars, or tumblers per minute. The key to this high-speed, close-tolerance registration process is the hot-melt medium that transports the glass enamel to the glass surface. Lower melting waxlike materials, into which the colors have been blended, are made fluid by electrical resistance heating of wire cloth screens until the squeegee delivers them through the image to the glass substrate. The color is immediately frozen by the cool glass in wrap-around fashion, if desired, enabling another color to be applied in rapid succession.

Spray application is another direct application method which is quite popular. Overall coverage, or masked partial coverage is possible. Sometimes intricate clear glass designs can be produced by screen printing a ceramic resist and spraying the entire surface. The resist, a wax containing inert fillers, "burns out" in the lehr, leaving a powdery residue which can be mechanically brushed or washed away.

Other direct methods include banding, dipping, stamping, and hand painting.

Pad-transfer printing has been used on china and whitewares for many years. Its use is becoming more prevalent on transparent glasses with the advent of silicone rubber transfer pads, which enable total transfer of enamel thicknesses having sufficient opacity to be attractive. Gravure plates, photoetched in metal or plastic, provide design areas that are flooded with hot melt or cold-color pastes and are then doctored clean except for the etched areas. The soft rubber pad picks up the wet image and moves it to and in contact with the glass substrate where it is deposited in total with no film split. This method affords the advantage of being able to print complex curves and textured surfaces not possible with other techniques. Machines are available that permit the use of common printing screens, which deliver a wet image to a flat plate prior to removal and transfer through the use of a soft pad.

Indirect Method

The most common method of indirect application is through the use of decals. Glass enamels in special vehicles allow for high pigment loading and low application thickness. Precision web printers with tight register controls permit multicolor designs being made which have a thin cross section even though several colors overlay one another.

Decals can be made in heat-release or water slide-off form. The heat-release type features a reverse-printed web with a heat-activated adhesive on the top surface which allows the colored image to selectively adhere to the substrate as a heated platen presses against the web and melts a wax on the carrier paper, disengaging the decal from the support web. Equipment employing stacks in magazines also have found use in heat-release decal decorating.

Water slide-off decals have a water-soluble adhesive between the colors and the support paper which dissolves in water, enabling one to slide the decal onto the glass article being decorated. Care must be taken to squeegee excess water and air from under the decal and to subsequently dry the decal several hours before firing. Both forms of decal can be fired in conventional lehrs.

New Materials

Enamelescent Colors

Ceramic glass enamels, having nacreous pigments added to achieve an entirely different lustrous sheen, were introduced in 1980. Because of a unique property involving light diffraction, it is possible to print an enamelescent color over a standard enamel and over bare glass to achieve three distinctly different colors. When used alone, one color will create two different colors, one in reflected light, the other with transmitted light. These new colors are available in spray, cold oil, and hot melt media. They are covered by U.S. Patent No. 4 353 911, issued October 14, 1982.

Ultraviolet Curing Vehicles

Ultraviolet curing organic inks have been available for some time. These resin vehicles are comprised of a variety of photoreactive resins, photoinitiators, and catalysts. Because glass enamels are generally 80–95% clear flux, light transmittance is fairly high when enamel powders are dispersed in UV vehicles. For this reason, thicker layers can be applied and cured than could be expected with opaque organic inks.

UV-cured systems are being used for printing automotive and architectural glass. A few seconds or less exposure to light sources of 200–300 W/in. (508–762 W/cm) harden the ceramic enamel inks for immediate handling — still cool to the touch.

Other applications include enamel inks and cover coats for making decals. This eliminates long air-drying cycles for multicolor decal preparation.

It is understood that some of these vehicles make excellent carriers for ceramic enamels used in pad-transfer printing with or without light exposure before firing.

7
Lead Frits

"Frit" is variously defined as compounded glass that is quenched and ground as a basis for glazes or enamels,[1] the small friable particles produced by quenching a molten glass material, or a glass that contains fluxing materials and is employed as a constituent in a glaze, body, or other ceramic composition.[2] That is, the term "frit" refers to materials, primarily oxides, which have been melted to the liquid state and then quenched to a glass and simultaneously broken up into particulate matter.

The fritting process begins in a manner similar to glassmaking. The required raw materials are usually technical grade oxides, hydrates, and/or carbonates. Small amounts of fluorides are sometimes used. These raw materials are weighed out and blended before smelting.

Frits are smelted by using gas, oil, or electricity in either boxlike or rotary-type smelters with refractory linings. The boxlike smelters (resembling small glass tanks) lend themselves to both continuous or batch processes, while the rotary-type smelter is batch operated. In the batch process, a charge of material is placed in the preheated smelter, brought to the desired temperature, and held for the time period necessary to achieve complete solution of the batch, typically 15–150 min. In the rotary-type smelter, the furnace is slowly rotated during smelting to enhance mixing and heat distribution. After smelting, the burners are shut off and the molten frit discharged and quenched. Rotary smelter capacity ranges from 5 kg in small laboratory smelters to 900 kg in production-size units.

In the continuous fritting process, raw material is continuously added to a pile at one end of the preheated boxlike smelter by a screw feeder or other means. The burners are directed at the pile, and molten material flows by gravity to the other end of the smelter where it is continuously discharged and quenched. Capacity typically ranges from 900 to 18 000 kg daily.

Pot or crucible smelting is rarely used for making frits except for (laboratory) experimental purposes or in the cases of high-purity glasses when platinum-lined crucibles are used to produce compositions of exacting tolerance.

Leaded frits are smelted at temperatures ranging from 980°C to greater than 1315°C to achieve complete reaction and a homogenous glass. After smelting, the molten glass is quenched and broken up, either by pouring it into water to produce a granular frit with a "popcorn"-like appearance or by passing it between water-cooled steel rolls and shattering the rolled sheet with a hammer to produce flake frit.

Frits are often used in this coarse quenched form to reduce dusting during later operations. Alternatively, they can be dry ball-milled to powder for greater ease of mixing in subsequent processing.

Why Are Lead Frits Used?

Adding lead oxide to a glaze or enamel by means of a frit serves several purposes.

In the first place, the fritting process is used to render lead oxide insoluble. Fritting lead into a glassy silica matrix provides an opportunity to control the acid solubility while combining the lead with other useful fluxes.

It should be stressed, however, that fritting alone does not guarantee a lower solubility. The acid solubility is directly related to composition and is not necessarily related to the quantity of lead that is present. For example, consider the two frits given in Table I. Frit A contains over 60% PbO and is insoluble in 4% acetic acid, the solution used in the standard ASTM lead release test.[3] On the other hand, frit B, which contains less than 10% PbO, is highly soluble in 4% acetic acid. Proper formulation for low lead release will be discussed below.

Fritting the lead also provides an opportunity to prereact the oxide with a variety of other raw materials. This prereaction simplifies the formulation and construction of a leaded glaze. It minimizes the subsequent reactions during the glost fire and helps to ensure a defect-free glaze. It, therefore, improves the rate at which the glaze matures in the glost fire. Premelting and prereacting also reduces the tendency to volatilize on remelting the glaze or other ceramic coating. Laboratory fired tests have shown no evidence of lead volatilization from leaded frits as high as cone 06 in all-fritted glazes.

Finally, fritting to an insoluble frit provides a convenient way to reduce the hazards of working with toxic lead oxide. This is particularly true when one is working with fritted material which has not been ground to powder. Such material has less surface area and is less dusty.

Types of Lead Frits

Lead can be used in the formulation of a wide variety of frit compositions. Not limited to simple lead bisilicates, lead is used for its fluxing action in complex alkali–alkaline earth–alumina–borosilicates.

Listed in Table II are several lead frit compositions[4] which provide the foundation for high-gloss, transparent glazes firing in the range of Orton cone 06 to cone 5.

Frits 1 and 2 are lead bisilicates and frit 3 a slightly calcium-modified lead bisilicate. Frits 4 and 5 are examples of low-melting borosilicate fluxes, illustrat-

Table I. Comparison of Lead Release from Two Selected Frits (Ref. 4) Used in a 99% Frit–1% Bentonite Glaze

Frit	Composition (wt%)							Pb release* (ppm)
	K_2O	Na_2O	CaO	PbO	B_2O_3	Al_2O_3	SiO_2	
A				61.3		3.1	35.6	1.16
B	5.3	9.0	6.3	6.5	20.3		52.7	141

*As per ASTM C-738 (Ref. 3).

Table II. Typical Lead-Containing Frits (from Ref. 4)

Frit	K$_2$O	Na$_2$O	CaO	MgO	BaO	ZnO	PbO	B$_2$O$_3$	Al$_2$O$_3$	SiO$_2$	ZrO$_2$	TiO$_2$	F
										Composition (wt%)			
1							65.0		1.0	34.0			
2							61.3		3.1	35.6			
3			5.6				67.6			26.8			
4						5.4	72.2	9.0		13.4			
5							43.2	21.9		34.9			
6							61.3		7.0	31.7			
7							50.0		20.0	30.0			
8		1.5					71.2		2.4	24.9			
9	2.5	1.2	8.1				33.7		5.2	49.3			
10	2.7	1.8	10.4				17.1	8.0	6.2	53.8			
11	1.1	2.8	6.9				3.5	24.2	5.6	55.9			
12		3.6	4.5				31.0	13.0	3.4	43.5	1.0		
13	0.3	2.7				7.6	43.5	5.0	5.0	35.9			
14	2.4	4.0	5.9			9.5	11.3	6.5	7.4	53.0			
15	1.9	1.8	5.5				28.4	13.0	3.8	40.8	1.2		3.6
16	4.1	2.8	3.3				30.8	6.1		43.9		9.0	1.3
17	2.5	1.6	9.6				15.8	7.4	5.7	52.2	5.2		
18	1.6	1.8	7.5	0.5			16.5	3.6	8.1	55.5	5.0		
19	2.3	3.8	5.5			8.8	10.6	6.2	6.9	47.5	8.4		
20	4.6	3.4	3.6				27.1	9.7	7.8	36.1		7.7	
21	2.2	0.6	6.5		15.0		30.9	12.3	3.7	29.8			

ing the use of two fluxes, lead and boron. On the other hand, frits 6 and 7 illustrate alumina additions to lead bisilicate, yielding frits of very low solubility. Frits 8 and 9 are more complex lead aluminosilicates, which can be used for all-fritted glazes. Frit 10 is a classic lead boroaluminosilicate suitable for all-fritted glazes firing in the range Orton cone 01 to 4. Frit 11 is a lower lead oxide frit useful primarily for its lower expansion. By contrast, frit 12 is a higher lead oxide containing material useful for lower-firing all-fritted glazes (cone 06–cone 1). Frit 13 is a low-melting, zinc-containing material. Frit 14 is an example of a zinc-containing material for higher firing *all-fritted glazes* (cone 2–4).

Frit 15 is an example of a fluorine-containing material for high-gloss, low-melting applications. Frit 16 is a titanium oxide containing material.

Frits are also available for preparation of opacified glazes. They lower the cost of such glazes by saturating the glaze's solubility for the opacifier (ZrSiO$_4$) using low-cost coarse smelter grade zircon as the ZrO$_2$ source, rather than the fine-particle higher-cost ZrSiO$_4$ opacifiers added at the mill. Thus, all the mill-added ZrSiO$_4$ is effective in opacification. Frits 17 and 18 are examples of zincless opacified frits, while frit 19 is an example of a zinc-containing material.

Frits can also be formulated to assist in the preparation of matte glazes. Frits 20 and 21 are examples of such materials.

Properties of Lead Frits

The variety of lead-containing frits available reflects the required range of physical properties needed by various ceramic coating applications.

In each case, the frit must fuse to a homogeneous, viscous glass at an appropriate temperature. In the case of lead-containing frits, this temperature is always below 1200°C, at which temperature lead oxide becomes highly volatile. For this reason, lead glazes are seldom used above cone 6.[5] The choice of temperature below 1200°C depends upon the available firing equipment, the substrate selection, and also on the trade-off between stability of colors used for decoration and durability of the glaze. In general, glazes fired at higher temperatures (~1100°C, or cones 2–4) have superior durability. On the other hand, several important ceramic colors are not stable above 1000°C. Glazes containing these pigments must generally be fired at cone 04 or below in specially formulated compositions.

The melting ranges of the various frits are a function of their composition. Table III gives the melting ranges of the frits listed in Table II. All of the various

Table III. Properties of Lead-Containing Frits (from Ref. 4)

Frit	Melting range (°C)	Coeff. of thermal exp. ($\times 10^{-6}/°C$)	PbO leachability per Currier test* (%)
1	715–790	6.8	0.3
2	760–815	6.7	0.1
3	675–845	8.1	85.8[†]
4	500–520	8.6	96.0[†]
5	795–815	5.2	93.6[†]
6	760–845	6.0	1.2
7	980–1120	4.7	
8	595–675	9.0	80.3*
9	815–950	6.4	0.03
10	865–910	6.5	0.2
11	840–915	5.2	14.9
12	730–790	6.5	3.2
13	705–760	6.8	3.3
14	815–871	7.2	0.44
15	715–770	8.3	11.8
16	730–790	9.5	0.13
17	815–910	6.2	0.13
18	915–980	5.8	
19	800–880	6.6	
20	900–955	9.5	0.41
21	750–815	8.6	86.8[†]

*Reference 12.
[†]Considerable care should be used when handling leaded frits with high acid solubilities to avoid ingestion or inhalation. Frits like these have lost favor in the industry since reformulation often allows the introduction of lead in a much less soluble fritted form.

monovalent and divalent oxides are effective in reducing the melting temperature of the frit.[6] In general, their fluxing power is as follows in decreasing order: $Li_2O > PbO > Na_2O > K_2O > BaO > CaO > SrO > MgO > ZnO$. In addition, replacement of silica by B_2O_3 decreases the melting temperature as does a decrease in the silica, zirconia, and alumina contents. Use of several fluxing ingredients in proper proportion is more effective in decreasing the melting temperature than use of any single oxide.[5]

Another important property of a frit is its coefficient of thermal expansion.[5] When used in a coating, expansion relates to the fit of the coating to the substrate. If tensile stresses are to be avoided, the expansion of the glaze must be slightly less than that of the body.

The coefficients of thermal expansion of the frits in Table II are given in Table III. The expansion of frits has been shown to be a function of composition. In Table IV are found two sets of expansion–composition factors, as taken from the literature. In general, those oxides that have the greatest fluxing power give the highest expansion. This is a major trade-off in the selection of a frit. The unique benefit of PbO is apparent. For a material of very high fluxing power, it is only moderate in its effect on expansion.

Since lead oxide is a toxic material, a most important property of lead-containing frits is to render them insoluble so that the lead oxide cannot be released to the environment. It has now been well established that lead release of a frit or a glaze correlates with acid resistance.[9-11]

Those frits which show excellent acid resistance also show low lead release. An accepted measure of lead oxide acid resistance of frits is the Currier test.[12] In Table III will be found Currier test results for most of the frits given in Table II. Strictly speaking, it is acid resistance divided by exposed surface area that is a frit property, a relation developed by Podmore[13] and now called the Podmore factor. The Currier test circumvents this problem by using a constant particle size fraction.

Table IV.   Table of Expansion Factors

Oxide	Winkelmann and Schott* $(\times 10^{-7})$	English and Turner[†] $(\times 10^{-7})$
SiO_2	0.8	0.15
Al_2O_3	5.0	0.42
K_2O	8.5	11.70
Na_2O	10.0	12.96
Li_2O	2.0	
CaO	5.0	4.89
BaO	3.0	4.20
MgO	0.1	1.35
ZnO	1.8	2.10
PbO	3.0	3.18

*Reference 7.
[†]Reference 8.

The construction of a low-solubility acid-resistant lead glaze is very dependent on composition. Much work has been done in establishing relationships between acid solubility and composition of the glaze.[9,10]

It has been shown that silica, alumina, zirconia, and similar ions, such as titania and tin oxide, are effective in lowering the lead release of a glaze.

$$\text{Good} = 2[Al_2O_3] + [SiO_2] + [ZrO_2] + [TiO_2] + [SnO_2] \quad (1)$$

The concentrations given in Eq. (1) are expressed in the empirical formula concentrations which are conventionally used in the whitewares industry. The factor 2 outside of the brackets arises from the fact that there are 2 equiv of Al ions in 1 mol of Al_2O_3.

It has also been shown that alkalies, alkaline earths, boron oxide, fluorine, phosphate, zinc oxide, cadmium oxide, and lead oxide are all fairly effective in increasing the lead release of a glaze.

$$\text{Bad} = 2([Li_2O] + [Na_2O] + [K_2O] + [B_2O_3] + [P_2O_5]) + [MgO] + [CaO] + [SrO] + [BaO] + [F] + [ZnO] + [PbO] \quad (2)$$

Again, the number 2 appears in order to relate the oxide concentrations to the ionic equivalents produced. Note that the "bad" elements consist of the soluble network former B_2O_3 and those ions such as F which disrupt the silicate network, together with all of the network modifiers.

These factors have been related to a data base of 77 glazes by using a computer-assisted regression analysis procedure. As a result of these studies, a figure of merit calculation was developed to predict in advance whether the lead release of a given glaze formula will be high or low. The best method of calculating the figure of merit was simply to divide the factor good [Eq. (1)] by the square root of the factor bad [Eq. (2)].

$$\text{FM} = [\text{Good}]/[\text{Bad}]^{1/2} \quad (3)$$

It was observed that when the figure of merit exceeded a value of 2.05, the lead release of the fired glaze was less than 7 ppm. On the other hand, when the figure of merit was less than 1.80, then some measurements of lead release exceeded 7 ppm, although individual readings may be less. The figure of merit was not able to discriminate acceptability of the lead release in the interval 1.80–2.05.

This figure of merit has been found to be applicable to predicting whether the lead release of a given formulation will be high or low, as is required for government regulatory purposes. A high degree of correlation with experimental data has been found, and the calculation is useful for sorting out acid-resistant glazes when new compositions are explored.

Only one qualification to this figure of merit has been noted. The figure of merit apparently does not apply to those cases where a soluble crystalline phase has been precipitated. For example, a glaze in which $PbTiO_3$ had precipitated had a lead release of more than 100 ppm, while its figure of merit was 2.31. This qualification limits the usefulness of the figure of merit to gloss glazes. Most opacified and matte glazes contain crystalline phases in addition to the vitreous matrix. In such cases, each phase must be considered separately, and that phase with the lowest acid resistance governs.

Applications

Leaded frits find application in glass enamels, porcelain enamels, and ceramic glazes. These subjects are treated in more detail in other parts of this book. Any lead used in the formulation of these coatings should be added as an acid-resistant lead frit.

References

[1]P. B. Gove, Ed.; Websters Third New International Dictionary of the English Language, Unabridged. G&C. Merriam Co., Springfield, MA, 1964.

[2]"Standard Definitions of Terms Relating to Ceramic Whitewares and Related Products"; 1982 ASTM Book of Standards, Part 17, ASTM C-242-77. American Society for Testing and Materials, Philadelphia, PA, 1982.

[3]"Standard Test Method for Lead and Cadmium Extracted from Glazed Ceramic Surfaces"; 1982 ASTM Book of Standards, Part 17, ASTM C-738-81. American Society for Testing and Materials, Philadelphia, PA, 1982.

[4]Pemco Technical Notebook on Ceramic Glazes and Stains. Pemco Ceramics Group, Mobay Chemical Corp., Baltimore, MD.

[5]F. Singer and W. L. German; Ceramic Glazes. Borax Consolidated, Ltd., London, England, 1964.

[6]P. Rado; An Introduction to the Technology of Pottery. Pergamon Press, Oxford, England, 1969.

[7]A. Winkelmann and O. Schott, "Uber Thermische Widerstandskoeffizienten Verschiedener Glaser in ihrer Abhangigkeit von Chemischen Zusammensetzung," *Ann. Phys.*, **51**, 730–46 (1894).

[8]S. English and W. E. S. Turner, "Relationship Between Chemical Composition and the Thermal Expansion of Glass," *J. Am. Ceram. Soc.*, **10** [8] 551–60 (1927); **12** [12] 760 (1929).

[9]R. A. Eppler, "Formulation of Glazes for Low Pb Release," *J. Am. Ceram. Soc.*, **54** [5] 496–9 (1975).

[10]R. A. Eppler, "Formulation and Processing of Ceramic Glazes for Low Lead Release," Proceedings of the International Conference on Ceramic Foodware Safety, Lead Industries Association, New York, 1975.

[11]J. E. Marquis, "Lead in Glazes—Benefits and Safety Precautions," *Am. Ceram. Soc. Bull.*, **50** [11] 921–3 (1971).

[12]E. R. Currier, "Standard Method for Determining Leachability of Lead Frits," *J. Am. Ceram. Soc.*, **30** [11] 335–8 (1947).

[13]H. L. Podmore, "The Specific Exposed Surface and the Lead Solubility of a Frit," *Trans. Br. Ceram. Soc.*, **50** [6] 262–4 (1951).

8
Electronics

Lead-based ceramics are critically important in electronic applications. As discussed in previous chapters, ceramic glazes and glasses containing lead are used in many parts of the electronic ceramic industry.

In this chapter, lead-based piezoelectric and ferroelectric material will be discussed in detail. In these materials, lead compounds are fundamental to the observed behavior. There are many other electronic ceramic formulations that include lead in the composition or as a flux. These, however, are too numerous to discuss in any detail and will not be covered.

Piezoelectricity and ferroelectricity are important classes of electrical behavior in materials. Piezoelectricity is defined as electric polarization produced by mechanical strain in some crystals, the polarization being proportional to the strain and changing sign with it.[1] Ferroelectricity is the spontaneous alignment of electric dipoles by their mutual interaction in some materials. The outstanding property of ferroelectrics is the reversibility of the permanent polarization by an electric field.[2] For a polycrystalline body to exhibit piezoelectricity, the material must also be a ferroelectric. Thus, in ceramics, ferroelectricity and piezoelectricity are interrelated.

Ferroelectricity is the dielectric analog to ferromagnetism. The name ferroelectricity is derived from this analogy and does not imply that iron is important in the phenomenon. Ferroelectricity in a material is a consequence of permanent electric dipoles resulting from its crystal structure. In a ferroelectric material there are small regions within each single-crystalline region in which all of these permanent dipoles are aligned in parallel for a net nonzero electric polarization. These regions are called domains. In a related class of materials, antiferroelectrics, the permanent electric dipoles are aligned in an antiparallel manner which cancels the individual dipole moments out for a net zero polarization.

Though the domains in a ferroelectric possess nonzero polarization, in the as-formed material the domains are randomly oriented, leading to an overall zero polarization in the sample. The dipole moments in the various domains can be aligned by an externally applied electric field, yielding a nonzero polarization in the sample.

The ferroelectric effect as mentioned above is a crystal structure effect. If the crystal structure of a material is such that it does not yield permanent dipoles, it will not exhibit ferroelectricity. In ferroelectric materials, the effect is observed only in those phases that have permanent dipoles. Thus, as any temperature or composition variations that will lead to phase changes occur in ferroelectric materials, ferroelectric behavior will be observed or not, depending upon the crystallographic phase present. The temperature below which ferroelectricity is observed is called the Curie point of the material. Above the Curie point, the material is nonferroelectric or paraelectric.

Piezoelectric materials convert mechanical energy to electrical energy and vice versa. The conversion of an applied mechanical stress (force) to an electrical polarization is called the direct piezoelectric effect. Conversion of an applied electrical voltage to a mechanical strain (deformation) is called the converse piezoelectric effect. A material that displays one effect will display the other. An important parameter of a piezoelectric is its electromechanical coupling factor, a measure of its conversion efficiency.

As with ferroelectricity, piezoelectricity is a consequence of the crystal structure of a material. The piezoelectric effect requires a noncentrosymmetric crystal structure. In materials possessing these structures, the application of a stress leads to electrical dipoles being created. In different grains of a polycrystalline piezoelectric material, these dipoles will point in different directions and end up canceling out. For this reason, it was thought until the 1940's that polycrystalline piezoelectrics were impossible. It was shown, however, that if a ferroelectric material is used and it is subjected to a high electric field to align the domains, a polycrystalline piezoelectric could be obtained. (This process of applying a high electric field is called poling.)

The first commercially useful piezoelectric ceramic was $BaTiO_3$.[3] Though $BaTiO_3$ was a useful piezoelectric and it had many applications, an important advance was the discovery of the piezoelectric properties of the $PbTiO_3$-$PbZrO_3$-PbO solid-solution series.[4] This solid-solution series, often designated PZT for lead zirconate titanate, has better piezoelectric properties than $BaTiO_3$ and can be used to a higher temperature (300 vs 120°C) than $BaTiO_3$. Table I lists some typical properties of the two types of materials.

Table I.   Typical Properties for Piezoelectric Ceramics*

Properties	Barium titanate	Lead zirconate titanate
Specific gravity	5.60	7.70
Dielectric constant at 25°C (Hz)	1200	1300
Dielectric strength (V/mil)	90	100
Modulus of elasticity (10^6 psi)	17	12
Piezoelectric constants d_{31}	-60×10^{-12}	-120×10^{-12}
d_{33}	150×10^{-12}	290×10^{-12}
g_{31}	-5×10^{-3}	-10×10^{-3}
g_{33}	14×10^{-3}	25×10^{-3}
Coupling coefficients K_{31}	0.2	0.3
K_{33}	0.5	0.7
K_p	0.3	0.6
Mechanical Q	400	500
Curie temp (°C)	120	345

*From Koenig (Ref. 5).

In this chapter, the PZT family of materials will be discussed. Related, yet importantly different, materials such as PLZT (lead–lanthanum–zirconate–titanate) and $PbNb_2O_6$ will also be discussed.

PbTiO$_3$

In 1950, it was reported that lead titanate was ferroelectric on the basis of indirect evidence.[6] It was not until 1970 that ferroelectricity in $PbTiO_3$ was actually verified.[7] Lead titanate has the cubic perovskite crystal structure above its Curie point of approximately 490°C. Below the Curie point the structure is tetragonal. At room temperature, the lattice constants are $a = 3.895$ and $c = 4.146$ Å for a c/a ratio of 1.064.[8] This relatively large tetragonal distortion of the cubic perovskite structure leads to a large amount of strain in cooling through the cubic–tetragonal transition temperature (490°C). This spontaneous strain and a large thermal expansion anisotropy is thought to be the cause of breaking up of polycrystalline bodies of pure $PbTiO_3$ during cooling.[9] For these reasons, polycrystalline $PbTiO_3$ has to be formed by using additives.[10]

$PbTiO_3$ is usually formed by the solid-state reaction of PbO and TiO_2. The compound can be formed at temperatures as low as 375°C.[11] Recently, however, a new technique for forming $PbTiO_3$ has been reported.[12] This technqiue involves preparation of a porous $PbTiO_3$ body from organometallic precursors at 70°C. The method offers the potential to lower the densification temperature of the material and may lead to fully dense, pure $PbTiO_3$.

Because of the difficulty in preparing pure polycrystalline lead titanate, there are few reports on the properties of the material. There are many reports on the properties of modified $PbTiO_3$.[13] However, the commercial application of these materials is still quite small.

PbZrO$_3$

Lead zirconate was shown by Roberts to have unusual dielectric properties in 1950.[14] Because of its high dielectric constant and temperature dependence, the material was designated ferroelectric. It was later shown that $PbZrO_3$ is actually antiferroelectric.[15] By itself, therefore, lead zirconate will not form a polycrystalline piezoelectric. Below approximately 230°C, $PbZrO_3$ is orthorhombic, with lattice constants $a = 5.87$, $b = 11.74$, and $c = 8.20$ Å. Above 230°C the material has a cubic crystal structure.

In his original work, Roberts prepared $PbZrO_3$ by mixing ZrO_2 and PbO and then calcining them in a platinum crucible for 1 h at 1050°C. As with the forming and densification of many lead-containing compounds, the volatility of PbO must be controlled. This is usually done by firing in an atmosphere containing PbO, as will be discussed in detail in the next section.

As with $PbTiO_3$, the properties of $PbZrO_3$ can be altered by modifying the composition. Some of the additions can change the crystallographic phases present in addition to altering the properties.[16]

Pb(Zr,Ti)O$_3$

Lead zirconate and lead titanate form a complete series of solid solutions.[17] These materials are referred to generically as PZT. It is important to realize that PZT refers to a wide range of compositions rather than an individual chemical compound.

The discovery of the piezoelectric properties of PZT ceramics was a result of research around the world. In 1952, Shirane and Suzuki, in Japan, investigated the crystal structure of the solid solution PbZrO$_3$-PbTiO$_3$.[18] This work was followed in 1953 by Sawaguchi's comprehensive report on his investigation of ferroelectricity in the PZT solid solution.[19] In 1954, Jaffe et al. from the United States published the first report on the piezoelectric properties of the PbZrO$_3$-PbTiO$_3$ solid solution.[4] These same authors summarized the work on PZT and expanded it in a more encompassing report in 1955.[20]

They showed that as Zr replaces Ti in PbTiO$_3$, the amount of tetragonal distortion (c/a ratio) decreases. At about 55 mol% PbZrO$_3$, an abrupt change in the crystal structure of the solid solution is observed. This sort of abrupt change in structure with compositional change is known as a morphotropic phase transformation. The morphotropic phase boundary (MPB) in PZT is nearly temperature independent. Solid solutions higher in PbZrO$_3$ content than 55 mol% have rhombohedral crystal structures. At very high PbZrO$_3$ content (>90 mol%), the crystal structure is orthorhombic. The tetragonal and rhombohedral phases are ferroelectric, while the orthorhombic phase is antiferroelectric.

PZT is commonly formed from the solid-state reaction of PbO, TiO$_2$, and ZrO$_2$. There have been many investigations into the calcination reactions occurring. In 1981, Chandratreya, Fulrath, and Pask described their work on the reaction mechanisms in the formation of PZT solid solutions.[21] Their results indicate that when the three oxide powders (PbO, TiO$_2$, and ZrO$_2$) are mixed and heated, first PbTiO$_3$ forms exothermically between 450 and 600°C. PZT then forms above 700°C. They were not able to observe any measurable amount of PbZrO$_3$ in the mixture. They proposed a multistep sequence to explain the reaction kinetics.

As with most technical ceramics, the processing of PZT articles is exacting and critical to their performance. The problems involved in balancing densification with grain growth is compounded in PZT by the volatility of PbO. Densification, grain growth, and evaporation rates all increase with increasing temperature. Thus, solely by varying the temperature during firing, it would be impossible to fully densify a PZT piece while minimizing grain growth and eliminating lead loss. In order to minimize lead loss, pieces are fired in a closed crucible containing a PbO-rich atmosphere. The PbO-rich atmosphere is usually created by packing the PZT pieces in a powder containing lead oxide. This packing powder can be of the same composition as the pieces or it could be PbO, PbZrO$_3$, etc. Chiang et al. reported on the effect of different packing powders on the microstructure and properties of PZT ceramics.[22] Though they did not perform the experiment, their results indicate that the ideal arrangement would be to use a closed system with a packing powder of PbZrO$_3$ saturated with PbO.

A major effort through the years has been devoted to preparing PZT ceramics with a good microstructure and good electrical properties. Hot-pressing has been used to obtain pieces with a dense, uniform microstructure. Kim and Hart have reported on their work on processing high-density PZT without hot-pressing by careful control of composition, impurities, calcining, forming, and sintering conditions.[23] It is common to use an excess of PbO in the PZT composition to act as a liquid-phase densification aid. Kingon and Clark report, however, that while a PbO-rich liquid phase increases the initial densification rate, it leads to a lower final density.[24] The answer may be, as pointed out by Fulrath, to remove the PbO liquid phase by controlled evaporation during the final stage of sintering.[25]

Jaffe et al. had found that compositions near the morphotropic phase boundary (MPB) yielded materials with desirable piezoelectric properties over a wide temperature range. The most useful commercial PZT compositions, therefore, are near the MPB. A commonly used composition has a Zr/Ti ratio of 53/47.

Soon after the discovery of the useful piezoelectric properties of PZT were demonstrated, work was directed to improving these properties via chemical modifications. Kulcsar reported in 1959 on the effects of partially substituting calcium or strontium for lead.[26] He found strontium (Sr) especially effective. When Sr is substituted for Pb in PZT, the dielectric constant and the electromechanical coupling factor increase over that of PZT near the MPB. Strontium substitutions were also found to lower the Curie point and shift the MPB to a slightly more Zr-rich composition. Kulcsar also investigated the effect of modifying PZT near the MPB with some three- or five-valent ions (La, Nd, Nb, and Ta).[27] These additions increased the dielectric constant and gave better aging and stability characteristics of the PZT. The same author found that modifying PZT with tungsten and thorium gave results similar to those containing Nb or La.[28] Weston et al. investigated the effect of additions of Fe_2O_3 to PZT compositions containing between 45 and 60 mol% $PbZrO_3$ (i.e., on both sides of the MPB).[29] The mechanical and dielectric stiffnesses were increased while the mechanical and dielectric losses were decreased by the addition of iron. The effects were greater for the rhombohedral than for the tetragonal ($PbTiO_3$-rich) compositions.

Solid solutions of PZT with other perovskite ferroelectrics have also been studied. Several of these have involved a material Pb $(Nb_{2/3}X_{1/3})O_3$ including those with X being Mg,[30] Co,[31] or Ni.[32] In these studies, it was found that the system contains pseudocubic as well as tetragonal and rhombohedral phases. The materials produced had high dielectric constants and electromechanical coupling coefficients and good temperature stability.

As described in the introduction to this chapter, piezoelectric materials convert mechanical energy to electrical energy and vice versa. There are many possible applications for devices based on piezoelectric materials. The PZT family of materials are the most commercially important ceramic piezoelectrics. By suitable modification of the composition and control of the microstructure through processing variations, materials can be tailored to the specific applications. The following discussion of applications is by no means complete but does give some indication of the important uses of PZT and the wide variety of those uses. Table II lists some of these applications.

Table II. Applications of PZT Ceramics

Application	Devices
High voltage generators	Igniter for gas appliances
	Cigarette lighter
	Small gasoline motor spark generator
	Static remover
Flexural element	Buzzer
	Liquid level sensor
	Fine motion controller
	Strain gage
	Phonograph pickup
Resonators	Electrical filters
	Electrical resonators
Sound/ultrasound generators	Remote control of appliances
	Intruder alarm
	Tone generator
	Echo sounder
	Hydrophone
	Ultrasonic cleaners
Electrooptics	Displays
	Flash shield

A useful monograph describing the applications of PZT ceramics is entitled "Piezoelectric Ceramics".[33] The monograph does not give details of the compositions used and uses only one company's (Philips:Mullard) materials as examples. It does, however, give the operating characteristics of the materials discussed and how to use them in a variety of specific applications. Much of the following discussion is based on this monograph. Banno has described some of his work on modifying PZT for several specific applications.[34] He found that W-modified PZT is suitable for piezoelectric ignition of gases, W–Mn-modified PZT is useful as ceramic wave filters, while Nb–Mn-modified PZT is suitable for high-power ultrasonic transducers. Newnham et al. have found that composites made from PZT and epoxy offer some unique properties which can be exploited in hydrophone and other applications.[35]

As shown in Table II, PZT can be used to generate high voltages. The voltages produced are high enough (5–20 kV) to cause a spark. The spark can ignite gas in an appliance such as a stove or clothes dryer. With a piezoelectric igniter, pilot flames are unnecessary, reducing the amount of gas used by the appliance. The mechanical force can be applied to the PZT piece (usually cylindrically shaped) by a spring-loaded hammer or by a lever system. The latter type of system has been used in a device which, when the lever is pressed, generates a high voltage, which ionizes air in the vicinity of the tip of a wire connected to the PZT block. The ionized air is used to remove static from phonograph records.

Flexural elements are also commonly made from PZT ceramics. The amount of motion generated by a single PZT element is often too small to be used (e.g., 4 μm for an applied voltage of 10 kV).[33] In these flexural applications, a laminated cantilever structure, or "bimorph," is used. A bimorph consists of two thin layers of PZT laminated together. Often, a metallic layer is used between the PZT layers to increase the durability of the bimorph and increase the electromechanical coupling. The way in which the bimorph cantilever is mounted affects the possible flexure modes and is a way to tailor the device to the application. Some of the uses of PZT as a flexural element are also given in Table II.

The resonance possible in PZT has been exploited in electrical circuits as resonators and filters. The dimensions of the PZT disk and the mode of oscillation (e.g., radial or thickness) initiated control the frequency of resonance of the device.

PZT transducers can be used to generate sound and ultrasound in air. The same materials can be used as receiver for these signals. Remote control of slide projectors and televisions have been developed on the basis of these systems. Other uses for this type of device are intruder alarms, telephone microphones, and tone generators. As with flexural elements, laminated structures are commonly used for ultrasound generators or receivers. A disk of PZT is often attached to a metal disk or a diaphragm. The mechanical design of the overall structure (i.e., size, clamping, etc.) determines many of the operating characteristics of the device.

Boats commonly have echo sounding equipment to determine the depth of the water and in some cases to locate fish. The systems used in small boats are usually designed with a maximum range of 100 m. This requires a PZT transducer with resonant frequency in the range 150–200 kHz. This transducer is also capable of receiving the signal. Depending on the specific application, the directionality of the system can be varied. A related device is a hydrophone, which is used for underwater detection by passively listening rather than generating and receiving a signal. As mentioned above, it has been found that PZT/epoxy composites are effective in hydrophones.

When high-intensity ultrasonic signals must be generated, such as in ultrasonic cleaning, a sandwich or composite transducer is used. The intensity of the signal required necessitates a large volume transducer. If it were to be made from a single PZT element, not only would manufacturing problems be enormous, but the large volume of the transducer would dissipate much of the energy.

PZT, when doped with rare earth additives, was found to act as an electrooptic material;[36] that is, the optical properties could be electrically controlled. This will be discussed in more detail in the next section on PLZT.

As can be understood from the above discussion, the PZT family of materials offers a wide range of properties. These properties can be adjusted by compositional and processing variations leading to materials with many useful applications.

PLZT

As discussed above, doped PZT has been found to be useful as an electrooptic material. An early composition used for this purpose contained 2 mol% Bi and a Zr/Ti ratio of 65/35.[37] It was then shown by Haertling and Land that lanthanum

(La) doping gave PZT ceramics with good electrooptical properties and a high optical transparency.[38] This family of materials is designated PLZT for lead–lanthanum–zirconate–titanate. A useful electrooptic composition has a Zr/Ti ratio of 65/35 and 8–9 mol% La. The specific compositions are often designated by a three-term ratio of the form La/Zr/Ti mol%, such as 9/65/35 PLZT.

The phase relations in the $X/65/35$ PLZT system with $0 \leq X \leq 15$ was investigated by O'Bryan.[39] It was shown that at room temperature two ferroelectric rhombohedral phases exist, with the transition occurring at about $X = 1$. At $9 \leq X \leq 12$, a pseudocubic phase was found. This phase was designated as antiferroelectric on the basis of dielectric properties. For compositions with $X > 12$, a cubic phase was identified.

In the initial work on PLZT by Haertling and Land, the samples were hot-pressed to reach optical transparency.[38] In the hot-pressing fabrication technique developed by Haertling,[40] the calcined powder was initially cold-pressed at 3500 psi (24.132 MPa) in an alumina hot-press mold. The mold assembly was then placed into the hot-pressing furnace. Pressure was applied from both ends. Typical hot-pressing conditions used to form PLZT devices were 1100°C for 16 h at 2000 psi[38] (13.79 MPa). When temperatures greater than 1100°C were used, Haertling used ZrO_2 powder as a setter to prevent reaction between the alumina mold and the PLZT. Hot-pressing is still used to fabricate PLZT pieces. Snow demonstrated a process for fabricating transparent PLZT ceramics without hot-pressing.[41] In this process cold-pressed 9/65/35 PLZT samples were first sintered in O_2 in platinum crucibles for 45 min at 1180°C. They were then heat treated in air for 60 h at 1200°C in Al_2O_3 crucibles containing $PbZrO_3$ powder.

Displays made from PLZT can image, store, and display information as can other display technologies. PLZT displays have several advantages over these types of electronic displays.[42] They can be selectively erased and do not need to be refreshed; that is, part or all of the information can be erased, and once information is input, no power is required to maintain it. The memory is nonvolatile.

The thin PLZT plate used in displays is transparent. Initially it is optically isotropic, but poling can reduce the symmetry to uniaxial. A uniaxial material is birefringent; that is, the index of refraction is different for different polarizations of light. By selectively reorienting the domains in the ceramic, the birefringence can be electrically controlled and an image stored. For a more detailed description of the operation of ferroelectric displays the papers by Meitzler and Maldonado[36,42] are very informative. In addition to information displays, electrooptic PLZT has been used for flash-protection goggles. In this application when a light sensor is triggered by a high-intensity flash, a signal is sent to the transparent PLZT, which reorients the domains and darkens the material, blocking the transmission of the light flash.

$PbNb_2O_6$

Lead metaniobate ($PbNb_2O_6$) was the first nonperovskite oxide ferroelectric discovered.[43] After this discovery by Goodman, other lead-based nonperovskite ferroelectrics have been identified[44] but will not be discussed here.

The Curie point of $PbNb_2O_6$ is 570°C.[43] Above this temperature, the para-electric phase has a tetragonal crystal structure which is similar to some alkali tungsten bronzes such as $K_{0.57}WO_3$.[45] The lattice constants of this phase were determined to be $a = 12.56$ and $c = 3.925$ Å.[46] Below the Curie point two phases have been observed, a rhombohedral nonferroelectric phase and the orthorhombic ferroelectric phase.[47] The lattice constants of the ferroelectric form are $a = 17.65$, $b = 17.91$, and $c = 7.736$ Å.[45] The ferroelectric phase is metastable at room temperature.

In his original work, Goodman prepared lead metaniobate in a two-step process.[43] The first step is a calcination of Nb_2O_5 and $PbSO_4$ in the form of pressed pellets at 1275°C for 1 h in air to form $PbNb_2O_6$. The calcined pellets were crushed and repressed. The final firing was carried out in platinum vessels at 1250°C. Lead volatility is not as much of a problem in forming lead niobate as it is in PZT.[48]

Modified lead niobate has been used in transducers. Due to its high Curie point, $PbNb_2O_6$ can be used at higher temperatures than PZT, but the relatively high electrical conductivity of lead metaniobate limits some of its use at higher temperatures.[48]

References

[1]W. G. Cady; p. 4 in Piezoelectricity. McGraw Hill, New York, 1946.
[2]W. Kanzig; p. 5 in Ferroelectrics and Antiferroelectrics. Academic Press, New York, 1957.
[3]B. Jaffe, W. R. Cook, Jr., and H. Jaffe; pp. 1–5 in Piezoelectric Ceramics. Academic Press, New York, 1971.
[4]B. Jaffe, R. S. Roth, and S. Marzullo, *J. Appl. Phys.*, **25**, 809–10 (1954).
[5]J. H. Koenig, "Edward Marburg Lecture," American Society for Testing and Materials, Philadelphia, PA, 1964.
[6] G. Shirane, S. Hoshino, and K. Suzuki, *J. Phys. Soc. Jpn.*, **5**, 453–5 (1950).
[7]J. P. Remeika and A. M. Glass, *Mater. Res. Bull.*, **5**, 37–45 (1970).
[8]T. Y. Tien, E. C. Subbarao, and J. Hrizo, *J. Am. Ceram. Soc.*, **45** [12] 572–5 (1962).
[9]T. Y. Tien and W. G. Carlson, *J. Am. Ceram. Soc.*, **45** [12] 567–71 (1962).
[10]E. C. Subbarao, *J. Am. Ceram. Soc.*, **43** [3] 119–122 (1960).
[11]S. S. Cole and H. Espenschied, *J. Phys. Chem.*, **41**, 445–51 (1937).
[12]S. Gurkovich and J. B. Blum; in Ultrastructure Processing of Ceramics, Glasses and Composites. Edited by L. L. Hench and D. R. Ulrich. J. Wiley, New York, 1983.
[13]B. Jaffe, W. R. Cook, Jr., and H. Jaffe; pp. 119–23 in Ref. 3.
[14]S. R. Roberts, *J. Am. Ceram. Soc.*, **33** [2] 63–66 (1950).
[15]E. Sawaguchi, H. Maniwa, and S. Hoshino, *Phys. Rev.*, **83**, 1078 (1951).
[16]B. Jaffe, W. R. Cook, Jr., and H. Jaffe; pp. 123–31 in Ref. 3.
[17]S. Fushimi and T. Ikeda, *J. Am. Ceram. Soc.*, **50** [3] 129–32. (1967).
[18]G. Shirane and K. Suzuki, *J. Phys. Soc. Jpn.*, **7**, 333 (1952).
[19]E. Sawaguchi, *J. Phys. Soc. Jpn.*, **8**, 615–29 (1953).
[20]B. Jaffe, R. S. Roth, and S. Marzullo, *J. Res. Natl. Bur. Stand.*, **55**, 239–54 (1955).
[21]S. S. Chandratreya, R. M. Fulrath, and J. A. Pask, *J. Am. Ceram. Soc.*, **64** [7] 422–5 (1981).
[22]S. S. Chiang, M. Nishioka, R. M. Fulrath, and J. A. Pask, *Am. Ceram. Soc. Bull.*, **60** [4] 484–9 (1981).
[23]Y. S. Kim and R. J. Hart; pp. 323–33 in Processing of Crystalline Ceramics. Edited by H. Palmour III, R. F. Davis, and T. M. Hare. Plenum, New York, 1978.
[24]A. I. Kingon and J. B. Clark, *J. Am. Ceram. Soc.*, **66** [4] 256–60 (1983).
[25]R. M. Fulrath; pp. 211–8 in Abstracts of the US–Japan Seminar on Basic Science of Ceramics, 1975.
[26]F. Kulcsar, *J. Am. Ceram. Soc.*, **42** [1] 49–51 (1959).
[27]F. Kulcsar, *J. Am. Ceram. Soc.*, **42** [7] 343–9 (1959).
[28]F. Kulcsar, *J. Am. Ceram. Soc.*, **48** [1] 54 (1965).
[29]T. B. Weston, A. H. Webster, and V. M. McNamara, *J. Am. Ceram. Soc.*, **52** [5] 253–7 (1969).
[30]H. Ouchi, K. Nagano, and S. Hayakawa, *J. Am. Ceram. Soc.*, **48** [12] 630–5 (1965).
[31]T. Kudo, T. Yazaki, F. Naito, and S. Sugaya, *J. Am. Ceram. Soc.*, **53** [6] 326–8 (1970).

[32]H. Banno, T. Tsunooka, and I. Shimano; pp. 339–44 in Proceedings of the 1st Meeting on Ferroelectric Materials and Their Applications, 1975.

[33]Piezoelectric Ceramics. Edited by J. van Randeraat and R. E. Setterington. Mullard, London, 1974.

[34]H. Banno, paper no. 2-FS-80P, presented at the Pacific Coast Regional Meeting of the American Ceramic Society, San Francisco, CA, Oct. 27, 1980.

[35]R. E. Newnham, L. J. Bowen, K. A. Klicker, and L. E. Cross, *Mater. Eng.*, **2**, 93–106 (1980).

[36]A. H. Meitzler, J. R. Maldonado, and D. B. Fraser, *Bell Syst. Tech. J.*, **49**, 953–67 (1970).

[37]C. E. Land and P. D. Thatcher, *Proc. IEEE,* **57**, 751–68 (1969).

[38]G. H. Haertling and C. E. Land, *J. Am. Ceram. Soc.*, **54** [1] 1–11 (1971).

[39]H. M. O'Bryan Jr., *J. Am. Ceram. Soc.*, **56** [7] 385–8 (1973).

[40]G. Haertling, *J. Am. Ceram. Soc.*, **49**, 113–18 (1966).

[41]G. S. Snow, *J. Am. Ceram. Soc.*, **56** [2] 91–6, [9] 479–80 (1973).

[42]A. H. Meitzler and J. R. Maldonado, *Electronics,* 34–39 (1971).

[43]G. Goodman, *J. Am. Ceram. Soc.*, **36** [11] 368–72 (1953).

[44]B. Jaffe, W. R. Cook, Jr., and H. Jaffe; pp. 217–22 in Ref. 3.

[45]M. H. Francombe and B. Lewis, *Acta Crystallogr.*, **11**, 696–703 (1958).

[46]R. S. Roth, *Acta Crystallogr.*, **10**, 437 (1957).

[47]M. H. Francombe, *Acta Crystallogr.*, **9**, 683 (1956).

[48]B. Jaffe, W. R. Cook, Jr., and H. Jaffe; pp. 215–16 in Ref. 3.

9

Glass-Bonded Mica

This section deals with the use of lead glasses in the manufacture of glass-bonded mica, a moldable and machinable ceramic material.

The use of lead glass in combination with mica made possible the development of one of the most versatile ceramic materials, called glass-bonded mica, the ceramoplastic material.

In the late 1910's, Percy B. Crossley, an Englishman, looking for new uses for the abundance of mica available in India developed a marriage between lead glasses and mica. This development created a moldable/machinable inorganic electrical insulating material with uniformly high insulating properties, which could be used at high temperatures. For the electric power transmission, generation, and usage fields, a new era was born. This new material allowed designers of electrical producing systems flexibility and miniaturization not previously available to them. The material was used mainly as an arc barrier where high electrical energy (kVA) was available. In later years, various combinations of lead glasses and micas increased the uses of glass-bonded mica (ceramoplastic) in areas such as communications, navigation, aeronautics, electronics, nuclear energy, aerospace, lasers, and many other high-technology industries.

Before the discovery of glass-bonded mica, a void existed between many of the properties of organic plastics and inorganic ceramics. Some examples of this void are as follows:

Plastics are readily machinable while ceramics are not.
Plastics can be molded to close tolerances whereas ceramics cannot.
Metal inserts are molded into plastics but not into ceramics.
Ceramics are stable materials but plastics are not.
Ceramics do not smoke and/or burn whereas plastics do.
Ceramics do not outgas while plastics do.
Ceramics temperature usability is approximately eight times higher than that of plastics.

Glass-bonded mica helped fill these differences between ceramics and plastics, thus the term "ceramoplastic."

The following will explain what glass-bonded mica is, how it is made, its properties, and many of its uses.

What Is Glass-Bonded Mica — The Ceramoplastic?

Glass-bonded mica is a mixture of lead glasses (fritted glasses) in powdered form mixed with powdered mica. This mixture is then cold-pressed into a preform shape. The preform is heated, allowing the glass to melt, and the mica then becomes the filler for the combined materials. After melting occurs, the heated preform can be molded under extremely high pressures either by transfer- or by compression-molding techniques. The final molded combined material, which is a thermoplastic, is known as glass-bonded mica — the ceramoplastic. Its form varies according to the mold and/or molding techniques used to produce the final desired product.

Techniques Used to Manufacture Glass-Bonded Mica

When glass-bonded mica was first developed, a combination of lead borate glasses and natural mica (see Table I) was used to make a compression-molded sheet which could be machined to a desired shape for use as an electrical insulating material. Through the years, requirements for this inorganic material with its unique insulating properties increased in number. However, design configurations demanded fabricating techniques not attainable by compression-molded techniques. Compression molding was the first method used to manufacture glass-bonded mica.

The Plastics Department of General Electric Co., a source of glass-bonded mica in its early days, developed the molding of glass-bonded mica by the transfer-molding process. This made possible parts with varied configurations, with or without metal inserts molded in and made to extremely close tolerances. Cost of parts previously machined were reduced, especially when large quantities were required.

Transfer molding also allowed design engineers to use an inorganic material in place of a plastic part when temperature and/or complete dimensional stability precluded the use of plastics.

Table I. Lead-Glass Combinations in Glass-Bonded Mica

Low-temperature glass-bonded mica formulas with continuous use to 750°F (400°C)

(1) $PbO \cdot B_2O_3 \cdot H_2O$ + Muscovite or phlogopite mica + Na_3AlF_6
(Lead borate Meta glass[†] + natural mica + cryolite*)

(2) $PbO \cdot B_2O_3 \cdot H_2O$ + Fluorphlogopite mica + Na_3AlF_6
(Lead borate Meta glass[†] + glass + synthetic mica + cryolite*)

High-temperature glass-bonded mica formulas with continuous use to 1300°F (704°C)

(3) $PbO \cdot SiO_2$ + Fluorphlogopite mica + Na_3AlF_6

(4) $PbO \cdot Al_2O_3 \cdot SiO_2$ + Fluorphlogopite mica + Na_3AlF_6
(Lead alumina silicate + synthetic mica + cryolite)

*Cryolite is used as a fluxing agent in order to increase glass flowability and wetting capability.
[†]A designation which signifies that no metallic inclusions are present.

Synthetic Mica Increases Temperature Range

For many years the maximum continuous operating temperature range for glass-bonded mica was 752°F (400°C). The limiting factor to produce a higher temperature material was the natural mica content. The maximum temperature capability of natural mica was well below that of the available high-temperature glasses. The hydroxyl ions (in water) within the natural mica flakes limited the temperature usability of natural mica.

During World War II, workers in Germany, aware of the strategic importance of mica in its use in electronic products and a possible loss of their source of supply, developed a technique of manufacturing synthetic mica (Table II). This development allowed the use of higher melting temperature lead glass in the glass-bonded mica formula. Then material usable to 1300°F (704°C) became available.

Using the elements found in natural mica, with the elimination of the impurities and the hydroxyl ions, workers in Germany made a superior mica flake with temperature capabilities to 2200°F (1204°C) before destruction.

Documents describing the process for manufacturing synthetic mica were confiscated by the United States Government. The Bureau of Mines duplicated the process with success and then made the information available to the U.S. industry.

Lead Glass Used in Glass-Bonded Mica

With the use of both natural and synthetic micas in glass-bonded mica formulas, many variations in temperatures and properties were available (see Table III). The use of lead glass was found to be as effective for use in the higher temperature glass-bonded mica material as in the lower temperature mixes.

Table II shows the lead–glass combinations used in the formulas of the various glass-bonded mica material grades.

In the manufacture of glass-bonded mica, it was found that surface tension and viscosity during the meltdown phase were extremely important in creating a strong, stress-free material. The lead glasses not only exhibited excellent qualities in these areas, but also wet the mica flakes, which had the effect of making the combined materials even stronger in tensile strength.

It has also been found that many variations in properties of the final material can be made available by varying the chemical formulation of the lead glass. This versatility has increased the applications for the use of glass-bonded mica.

Why Glass-Bonded Mica Is Such an Important Electrical Insulation Material

Here is an inorganic material that can be formed into a sheet, a rod, or specific molded configuration. It is machinable on standard equipment to close tolerance. It is nonporous and does not absorb moisture; thus, its surface electrical resistivity is high. It does not *carbon track* during electrical arc-over. Before a restrike arc-over can occur, the current must be at the same potential as the first arc-over.

As more sophisticated applications appeared, it was found that glass-bonded mica could be molded to extremely close tolerances; i.e., the center-to-center

Table II. Synthetic Mica Formula

Name — Fluorphlogopite
$KMg_3AlSi_3O_{10}F_2$
Fluorine was added to replace the hydroxyl ion which is found in natural mica
Manufacturing process — resistance melting process
Other synthetic micas available — boron-, lithium-, and barium-containing micas

distances between molded-in metal inserts are held to ±0.005 in. (0.127 mm) when distance is 5 in. (127 mm). Flatness of 0.006 in. (0.152 mm) over 12 in. (304.8 mm) has been obtained. Outstanding capabilities such as these made this material desirable over other inorganic materials, which require a ±10% tolerance allowance.

Outstanding Properties of Glass-Bonded Mica

As previously explained, glass-bonded mica fills the properties gap between ceramics and plastics. Thus, we now have an inorganic high-temperature material that molds and can be machined as a plastic, although it is stable and acts as a ceramic.

Dimensional Stability

Glass-bonded mica maintains its dimensions even after continued thermal cycling. When the material is subjected to the maximum continuous operating temperature, it will change dimension due to its coefficient of expansion. When it is brought back to room temperature, the dimensions will be the same as they were before thermal cycling. Its stability is further exalted by the ability to place a molded part back into the mold cavity in which it was manufactured when both items are at room temperature. This feat can be accomplished both before and after thermal cycling.

Thermal Endurance

Glass-bonded mica has been thermal shocked by placing the high-temperature grade material into an oven at 1100°F (593°C) for 4 h. Upon removal from the oven, the part was dropped into a bucket of cold water at 50°F (10°C). When taken out of the water, examination of the part showed that it did not crack, chip, spall, or warp. A dimensional check showed no changes.

Arc Resistance

As previously stated, glass-bonded mica does not carbonize or carbon track due to electrical arc-over conditions; therefore, a restrike electrical arc-over condition would have to be applied at the same potential as that of the first arc-over.

Moisture Absorption

There is virtually no moisture absorption. A minute quantity of moisture can be trapped under flakes of mica which may have open edges on the surface of the part. This absorbed moisture will not penetrate more than approximately 0.003 in. (0.076 mm). A weight test before and after a part is put into water will not pick up this small amount of moisture. The moisture will quickly evaporate by a slight rise in temperature.

Table III. High-Frequency Effects on Glass-Bonded Mica

Material	Frequency (MHz)	Dielectric constant	Loss tangent (dissipation factor)
Natural mica	1	7.06	0.0014
	1000	7.33	0.0018
	3000	7.32	0.0014
	8500	7.54	0.0022
Synthetic mica	1	7.70	0.0014
	1000	7.65	0.0013
	3000	7.66	0.0010
	8500	7.67	0.0016

Thermal Conductivity

Glass-bonded mica is an excellent heat insulator. It takes twice as long for heat to pass through glass-bonded mica as it does through an asbestos-filled material. With respect to ceramic materials, it is approximately 8–10 times better as a heat insulator.

Electrical Properties

It is an excellent electrical insulator. For high-voltage applications, it has been used in areas requiring electrical insulation of not less than 60 kV rms. Because of its low loss tangent (dissipation factor), it is used in many high-frequency applications (see Table III). Its volume resistivity at various temperatures can be found in Table IV. Note that even at high temperatures, its volume resistivity shows very little change.

Flame and Smoke Retardant

This material does not burn or smoke. At very high temperatures the material bloats and breaks apart.

Radiation Effects on Glass-Bonded Mica

There have been many tests performed on glass-bonded mica for use in radiation areas. Obtaining copies of the data has been difficult. Some facts known about its use in a radiation environment are as follows:

Glass-bonded natural mica is used as a shield to prevent radiation from escaping to an area where it can do damage. This is because of the high lead content within the material.

Glass-bonded synthetic mica is used within the radiation area. It is used as a bearing housing and other support structures within a cyclotron to act as an electric insulator and /or mechanical support. The use of a low-boron lead glass is advantageous in this application.

Type of Radiation

The type of radiation is very important in evaluating radiation effects. In a reactor, for example, important types of radiation that strike at materials not in

Table IV. Temperature vs Volume Resistivity at a DC Voltage of 500 V

Temp		Volume resistivity (Ω·cm)	
(°F)	(°C)	Natural mica	Synthetic mica
100	37.8	10^{12}	10^{13}
200	93.0	10^{13}	10^{14}
300	149.0	10^{12}	10^{13}
400	204.0	10^{11}	10^{12}
500	260.0	10^{10}	10^{11}
600	316.0	10^{8}	10^{9}
800	371.0	10^{5}	10^{8}
1100	593.0	material becomes soft at this point	10^{6}

contact with fuel are thermal neutrons, epithermal neutrons (all energies above thermal), and gamma rays.

For most ceramics, the neutrons cause change within the ceramic structure. The energy level of a neutron is described as neutron flux. This property is important because radiation-damage reactions occur at different neutron energy levels.

It requires neutrons of 1000 eV (electron volts) or more to displace a metal atom from its lattice to produce an interstitial vacancy within the metal structure. It takes approximately 100 eV to cause lattice expansion in a ceramic material. It requires only a few electron volts to break down a chemical bond in plastics.

Although the neutron flux type of radiation is usually the cause of breakdown in an organic material, the gamma-ray radiation also causes degradation.

Glass-bonded synthetic mica is made with a lead–alumina–silicate glass containing 7–15% alumina. This type of glass content allows this material to be used in a radiation environment — up to 800°C and 5.2×10^{16} neutrons/cm^2 (10 keV) (fast neutron) exposure and 3.4×10^{12} rads gamma-ray exposure.

Information on radiation resistance of glass-bonded mica was received from the following:

1. Brookhaven Laboratory
   Upton, NY
2. Nevins Cyclotron Laboratories
   Division of Columbia University
   Irvington-on-Hudson, NY
3. Norden
   Division of United Aircraft
   Norwalk, CT
4. Los Alamos Scientific Laboratory
   Los Alamos, NM
   (Used on original Van de Graff generator)
5. Canel
   Atomic Station
   Middletown, CT

6. Oak Ridge National Laboratory
   Oak Ridge, TN

Low-Level Outgassing

Glass-bonded synthetic mica material is by far the best grade of glass-bonded mica to use in an environment requiring low-level outgassing materials. The lead glass causes no problem as long as the temperature is below the specified continuous operating temperature of the material. The natural mica, however, which contains many impurities and hydroxyl ions, does create some instability under a hard vacuum. Thus, the mixture of pure synthetic mica and glass is the best choice.

The results of an outgassing test performed on glass-bonded synthetic mica were as follows: The material was placed in a vacuum chamber at 400°C. At the end of 1 h under vacuum and heat, 0.4 μL of gas evolved. Analysis showed that this gas contained 44% CO_2, 2% O_2, 2% CO, 5% N, 39% H_2O, and 8% H_2. Then heat was turned off in the vacuum chamber. At the end of 1 h at room temperature, 0.2 μL of gas evolved. This contained 9% CO_2, 1% O_2, 5% CO, 3% N_2, 75% H_2O, and 6% H_2. The temperature was then raised to 600°C in the vacuum chamber in order to measure residual gases. The 0.2 μL of gas which evolved contained 3% CO_2, 1% O_2, 3% CO, 3% N_2, 26% H_2O, and 64% H_2. Outgassing analysis of lead alumina silicate and synthetic mica material showed that initially 10.1 μL of gas evolved. At the end of 1 h under vacuum and heat 0.10 μL of gas evolved. Analysis of the 10.1 μL of gas showed that 11% CO_2, 1% CO, 3% N_2, 83% H_2O, and 1% H_2 were present.

The H_2O and H_2 are probably background gases that are present in the vacuum chamber, with slight traces possible from the surface of the glass-bonded mica sample. The CO_2 evolved from the mold release, which is a necessary part of the molding cycle in the manufacture of glass-bonded mica. The other gases may also evolve from the mold release. We do not feel that the glass used in the material will outgas at the elevated temperatures. When the glass-bonded synthetic mica was subjected to 800°F (427°C) under a vacuum of 1 $\times$ 10^{-9} torr (1 $\times$ 10^{-7} Pa), no appreciable outgassing was detected after 1000 h at this condition. When the synthetic mica was subjected to 1000°F (538°C) under a vacuum of 1 $\times$ 10^{-4} torr (1 $\times$ 10^{-2} Pa), no appreciable outgassing was detected after 72 h at this condition.

Machining Glass-Bonded Mica

This material is quite easy to machine with carbide tools, much water (containing a water-soluble oil as a rust inhibitor), and standard available machine shop equipment. Before machining, it is important that an appreciation of the material be gained, as it is a ceramic material which is brittle. Also, the sheet material used for machined parts is a lamina structure, due to the flat orientation of the mica flakes during compression molding. The strongest mode of the material is perpendicular to the length of the sheet; thus, consideration must be given to this phenomenon before committing a part to machining. Rods are made from the sheet material and then centerless ground to size as this material is not conducive to extrusion manufacturing techniques. Also note that the laminar structure is now parallel to the length of the rod. When machining this material, make sure that a sharp cutting tool is always used. Support the material adequately and use plenty of water. Some of the basic machining procedures are detailed here.

Grinding

It is best to grind this material whenever possible. Use any normally available bonded wheels (such as silicon carbide resinoid bonded) and use grinding speed recommended by the manufacturer. Cylindrical, flat, or grinding between centers are all approved procedures. For heavy grinding, use a soft coarse wheel, and for fine grinding or polishing, use a hard fine wheel. Centerless grinding is used when an extremely close tolerance outer diameter is necessary. The use of water as a lubricant/coolant helps obtain a smooth surface as well as close tolerances.

Cutting

Wheel cutting with a good bonded diamond cutoff wheel is the most satisfactory method of cutting glass-bonded mica. If a diamond wheel is not available, use a bonded silicon carbide type abrasive cutoff wheel, with plenty of water in the cutting area. Wheel speed should be about 2500 rpm. Cut down into the work. Work slowly and steadily. Do not force cutting to increase cutting speed.

A band saw with silicon carbide teeth can also be used. Cut slowly with this method; however, a high-speed saw blade with a 14-teeth set should be used. Cut at the rate of approximately 75 ft/min (23 m/min). No water is required with this type of cutting.

Flycutting

Use tungsten carbide cutters at slow speed. Cut half-way through the part and then turn the part over and cut from the other side until part is cut through. Use a ¼ in. (0.64 cm) pilot hole in center, if possible.

Drilling

Carbide drills, with cutting angle of 90° inclusive, will give the best results. Either a fluted or flat drill can be used with plenty of water as the coolant and lubricant. See Table V for drill speed information. High-speed steel drills can also

Table V.   Drill Speed

Drill diam (in.)	Decimal equiv (in.)	Decimal equiv (mm)	Drill speed* (rpm)
¹⁄₁₆	0.0625	1.588	1500
⅛	0.125	3.175	1000
¼	0.25	6.350	400
⅜	0.375	9.525	325
½	0.5	12.700	275
⅝	0.625	15.875	250
¾	0.75	19.050	200
⅞	0.875	22.225	175
1	0.875	25.400	150

*Drill speeds for best results, assuming a sharp drill is used.

be used; however, they must be sharpened very often. For small holes, high-speed drilling works well. Drill holes by using a pecking motion so that the debris can be washed out of the hole. Try not to drill the hole completely through the part. For the cleanest hole, drill until the drill point breaks through the material, and then turn the part over and drill from the other side. If you must drill from one side, support the bottom of the part where the hole will break through; otherwise, major chipping will occur at the exit end. Keep drills sharp and use plenty of water.

Slotting

This operation may be accomplished in several ways: Use a metal-bonded diamond wheel or silicon carbide wheel of required cutting width on a cutting grinder or milling machine. Use a tungsten carbide tool bit on a shaper or milling machine.

With cuts of around 0.005 in. (0.127 mm) deep, one can proceed rapidly. If deeper cuts are desired, reduce cutting speed to where chipping is not in evidence on the sides of the machine groove. Again, use plenty of water as a lubricant and coolant.

Turning and Threading

The use of a tool post grinder on a lathe gives the best finish and the fastest method of machining glass-bonded mica by turning. Use silicon carbide or metal-bonded diamond wheels.

Threading should be accomplished in the same manner; only use a formed wheel. Single-point carbide-tipped tool bits can be used; however, the finish is poor and some chipping usually occurs on the surface of the part.

Tapping

High-speed steel taps can be used. For long runs, carbide may be more economical as carbide taps last longer. Drill a hole approximately 0.005 in. (0.127 mm) larger than the recommended hole size for the tap being used (this eliminates a sharp minor diameter), countersink the hole slightly to provide a lead-in and then start to tap, using plenty of lubricant (water) to keep the tool cool.

Milling

This operation can be accomplished with either high-speed steel or carbide-end mills. The latter is more economical for long runs and also cuts at about twice the speed.

Spindle speed of around 50 rpm gives good results with minimal chipping. A round- or flat-end mill tool can be used with no restrictions on the number of flutes.

Surface Finishing

The material can be filed, sanded, or lapped quite easily. Any of these methods may be used to break sharp corners, repair a rough surface, develop a polished surface, and blend uneven areas. Silicon carbide waterproof sandpaper works the best.

Lubricant/Coolant

Enough emphasis cannot be put on the use of plenty of water, which contains a water-soluble oil, for use as a lubricant. This keeps the tool from getting hot as well as acting as a cutting lubricant.

Designing with Glass-Bonded Mica

The uniqueness of glass-bonded mica, as a ceramic material, is that besides its ability to be machined, it can also be molded.

Many varieties of glasses were tried in the molding of glass-bonded mica; however, the lead glasses were the most successful for use as a universal molding compound. Material flow was excellent. It filled out the mold cavities completely with consistency. Its coefficient of expansion allowed for easy molding in of metal inserts with varied expansion rates. These features allow a designer the freedom to design parts almost as he would a plastic part, except that he can have the properties of ceramics (see Table VI).

There are various design parameters that must be considered when designing a part for molding from glass-bonded mica (see Fig. 1).

1. This material has practically no shrinkage during the molding cycle; thus, any surface perpendicular to the opening of the press must have a draft angle to facilitate removing the part from the mold.
2. Use maximum radii wherever possible, except on the parting line.
3. Use a consistent wall thickness throughout the part for best results. The lead glass content in the material allows parts to be molded with wall thickness variations of 5 to 1, and in some cases as much as 10 to 1. However, for the best results, stick to a uniform wall thickness.
4. Insert molding requires a mechanical bond between the insert and the glass-bonded mica material; therefore, the insert should be knurled or undercut in the molding area.
5. Remember the material is machinable, so that additional configurations or closer tolerances can be machined into the part after molding.
6. If at all practical, use threaded molded-in metal inserts instead of molded-in inner or outer diameter threads. Threads can be molded directly into the material; however, the threads are quite weak; thus, using them more than once is difficult.
7. Since glass-bonded mica is a stable inorganic material, it can be designed so that it is under compression. It will not move or flow under this condition.
8. Unlike other materials, glass-bonded mica can be molded to close tolerances even if the part is as large as 5 in. (12.7 cm) long. This size part can still maintain ±0.005 in. (0.127 mm) and ±0.003 in. (0.076 mm) in specific dimensions within the cavity boundary. Once the dimension is obtained within the part, it will not change even after continuous thermal cycling.

Insert material should be chosen so that its coefficient of expansion is close to that of the glass-bonded mica (see Table VII for the coefficients of expansion of different materials). Materials such as the 300 and 400 stainless steels give best results. Brass, copper, silver, and aluminum can also be molded in as they are softer metals and put little stress at the metal–glass-bonded mica interface.

Other metals such as titanium, nickel–iron blends, Dumet, and invar have also been molded into glass-bonded mica.

Table VI. Properties Chart for Glass-Bonded Mica

General properties	ASTM method	Natural mica		Synthetic mica	
		Transfer molded	Compression molded	Transfer molded	Compression molded
		Typical data only			
Specific gravity (g/cc)	C373	3.8	3.0	3.9	2.8
Density at 25°C (lbs/in.^{3} (g/cm^3))		0.14 (3.87)	0.11 (3.04)	0.14 (3.87)	0.10 (2.77)
Thermal conductivity ($\times$10) (cal $\cdot$ s^{-1} $\cdot$ cm^2/°C $\cdot$ cm^{-1})		12.0	10.0	12.5	14.0
Moisture absorption		Nil	Nil	Nil	Nil
Coefficient of thermal expansion ($\times$10^{-6} (in. $\cdot$ in.$^{-1}$/°C))		11.2	10.5	10.3	9.4
Specific heat (cal $\cdot$ g/°C)		0.24	0.12	0.23	0.11
Max. continuous operating temp. (°F)		750 400	750 400	1300 700	1100 538
Flammability		Does not burn			
Radiation resistance (3$\times$10^{10} Rad of Cobalt)		Good	Fair	Excellent	Excellent
		Electrical properties			
Dielectric strength (V/mil) ⅛ in. thick)	D149	375	400	375	380
Arc resistance (s)	D495	327	300	350	325
Permittivity (MHz)	D150	9.3 (8.4)	617	9.5 (8.8)	6.8
Dissipation factor (MHz)	D150	0.0012	0.0018	0.0015	0.0017
Loss index (MHz)	D150	0.0115	0.0120	0.0150	0.0120
Surface resistivity (dry Ω $\cdot$ cm) (70°F)	D257	10^{15}	10^{16}	10^{15}	10^{16}
Volume resistivity (Ω $\cdot$ cm) (70°F)	D257	10^{13}	10^{12}	10^{14}	10^{14}
Surface resistivity (wet Ω $\cdot$ cm) (70°F)	D257	10^9	10^6	10^{10}	10^{11}
Dielectric constant (MHz)	D150	9.3	6.7	9.0	6.8
		Mechanical properties			
Tensile strength (psi (N/mm^2))	D651	6500 (45.28)	6000 (41.80)	6000 (41.80)	5000 (34.83)
Flexural strength (psi (N/mm^2))	D790	12 000 (83.60)	15 000 (105)	10 000 (69.67)	12 000 (83.60)
Compressive strength	D695	33 000 (229.90)	45 000 (313.50)	30 000 (209)	32 000 (222.33)
Modulus of elasticity (in tension) (psi (N/mm^2))		7.0 (0.049)	11.0 (0.077)	9.0 (0.063)	10.6 (0.074)
Hardness — Rockwell H (H)		90	90	90	90
Impact strength — IZOD (notched) (ft-lbs/in.)		0.7	1.8	0.6	1.3
Poisons ratio (25°C)		0.26	0.62	0.26	0.57

*Glass-bonded mica summary specification — ASTM-D1039
†Conforms to United States military specification — MIL-I-10 Grade L4.

	Commercial	Special	Custom
Thickness tolerance	±0.005	±0.003	±0.001
Pin draft	1°	0.5°	0.25°
Side draft or irregular core	2°	1°	0.5°
Hole location tolerance	±0.003	±0.0015	±0.0008
Hole tolerance (either end as specified)	±0.003	±0.002	±0.001
Wall thickness (min.)	0.063	0.031	0.020
Fillets & rounds (min. desirable)	0.031	0.016	0.005
Gate projection	as molded	sanded	ground

Note: 1° = 0.017 in. per in.

Molded-in Inserts

Draft Angle and Radii

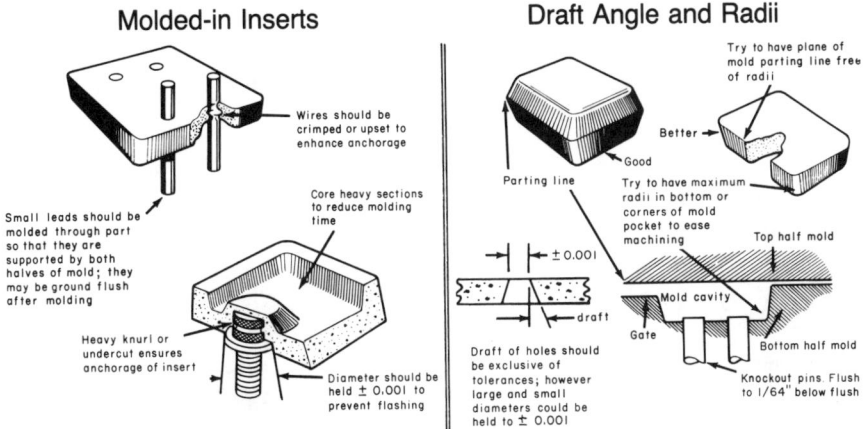

Wires should be crimped or upset to enhance anchorage

Small leads should be molded through part so that they are supported by both halves of mold; they may be ground flush after molding

Core heavy sections to reduce molding time

Heavy knurl or undercut ensures anchorage of insert

Diameter should be held ± 0.001 to prevent flashing

Try to have plane of mold parting line free of radii

Parting line

Better

Good

Try to have maximum radii in bottom or corners of mold pocket to ease machining

Top half mold

Mold cavity

±0.001

draft

Draft of holes should be exclusive of tolerances; however large and small diameters could be held to ± 0.001

Gate

Bottom half mold

Knockout pins. Flush to 1/64" below flush

Fig. 1. Molding design data.

Applications

Through the years, the range of applications for the use of glass-bonded mica has varied from a simple insulator washer to a complex molded part for lasers. Some of the many applications in which this family of materials have been used are given in Table VIII.

Health and Safety

The lead in the glass used in glass-bonded mica is a cause for concern in both environmental pollution and health. The lead in the final product is in a form that may leach into water. The waste from glass-bonded mica is considered a hazardous waste under the Resource Conservation and Recovery Act (RCRA) and must be handled according to the following: Federal Register — May 19, 1980; "Hazardous Waste and Consolidated Permit Regulations," 40 CFR Parts 260, 261, and 262, Environmental Protection Agency.

Air pollution via lead emissions from an industrial source is covered under 40CFR51 (App. B"3.4"), 40CFR51 (App. C), and 40CFR52.21 (b) (1). Water pollution from a point source category is covered under 40CFR124 (App. D,9).

Table VII. Linear Coefficient of Thermal Expansion (α) of Various Materials

Material	α ($\times 10^{-6}$/°F)	α ($\times 10^{-6}$/°C)
Silica glass	0.2	0.36
Nickel, cobalt, iron	0.9	1.6
Tungsten	2.2	4.0
Nickel, iron	3.0	5.4
Molybdenum	3.0	5.4
Platinum	4.9	8.8
Glass-bonded synthetic mica (compression-molded)	5.2	9.4
Nickel iron	5.4	9.7
Stainless steel (403, 410, 416)	5.5	9.9
Glass-bonded natural mica (compression-molded)	5.5	9.9
Glass-bonded synthetic mica (transfer molded)	6.0	10.8
Glass-bonded natural mica (transfer molded)	6.5	11.7
Monel	7.8	13.9
Gold	7.9	14.2
Steels (SISI-C1010-113)	8.4	15.1
Inconel X	9.0	16.2
Stainless steel (301 & 321)	9.5	17.1
Copper & red brass	10.4	18.7
Silver	10.9	19.6
Aluminum	13.9	25.0
Epoxies	38.3	68.9
Polytetrafluoroethylene	55.00	99.0

The generation of dust during the manufacture and machining of glass-bonded mica is to be avoided. Lead in the work environment is covered by the Occupational Safety and Health Administration (OSHA) under 29CFR1910. 1000.

The safe handling of glass-bonded mica and compliance to the applicable regulations cannot be overemphasized.

Manufacture: Dust collectors must be provided so that vacuum air ducts can be placed in areas where dust can be generated. Use covered or completely enclosed equipment wherever possible. In areas where some dust cannot be avoided, a NIOSH-approved dust mask should be worn.

Machining: Water is used to machine glass-bonded mica; thus, the waste water contains waste particles of the glass-bonded mica material. This waste should be collected, stored, and then disposed of in a manner specified by the Environmental Protection Agency (EPA). Disposal is generally accomplished by dumping the material into an approved landfill area.

Personnel Protection: It is recommended that a rigid hygiene training program be developed in order to assure that the people handling this product wash as often as possible, especially before handling and eating food, their work clothing is changed daily, and air is not used to blow off any dust from their clothing. (Uniforms should be supplied, if at all practical.)

Table VIII. Applications of Glass-Bonded Mica

High-voltage applications

Power stations	High-voltage electron tube bases
Tubes 7 sockets	Relay spacers, arc chutes 7 headers
Electric railroad cars	Commutators
Induction heating equipment	Transformers
Capacitor end plates	Receivers
Microwave components	Terminal boards
Heater coil supports	Coil forms
Connectors	Capacitors
Feed throughs	Switch gear
Stand offs	Computer components
Circuit breakers (baffle plates)	Transmitters

High-temperature applications

Jet engine thermocouple	Supports in furnaces to 1100°F temperature
Coil forms and bobbins	
Plasma engines	Heat barrier requirements
Thermal insulators	Spacers and washers
Brush holders	Connectors
Infrared components	Thermostat housings
Soldering fixtures	Mold platen precision spacers
Lightning arrestor insulators	Motor insulating spacers
Hermetic seal feed-throughs	Printed circuits
Plasma arc engines	Lamp brackets & housings
High-temperature instrumentation	High-temperature conveyor slats
Jigs & fixtures for induction heating	High-temperature terminals blocks
Heating element electrical feed-through	Switches

Dimensional-stability applications

Computer disks	Potentiometers
Semiconductor test fixtures	Accelerometer
Inspection fixtures	Memory boards
Mechanic supports	Encoding disks
Encoding memory devices	Laser components
Printed circuit boards	Gyroscopes
Antenna insulators	Telemetering devices
Commutation devices	Tunable oscillators
Aircraft instrumentation	Radar insulators
Sub bases for magnetic memories	

Other applications

X-ray equipment, fixtures and components	Vacuum pumps
	Antenna insulators
Radiation areas	Corona eliminators
Space exploration field	Cryogenic devices

Bibliography

J. Harry DuBois and Frederick W. John; Plastics. Van Nostrand Reinhold Co., New York.

"Mycalex — Electrical and Electronic Insulation," technical brochure, Mycalex Corp. of America, Division of Spaulding Fibre Co., Inc.

"Total Dimensional Stability," technical brochure, Mykroy Ceramics.

10
Glass-Bonded Abrasive Wheels

The manufacture of abrasive wheels is a modern process as we know it today. The first record of a successful vitrified grinding wheel being made in the United States was in 1869 by Sven Poulson,[1] a Swedish immigrant working for a small pottery shop owned by Frank Norton in Worcester, MA. As far as is known, there are no indications of the temperatures used. Since it was a pottery operation, one could assume it was a relatively low temperature, perhaps in the range of cone 1–5. There also is no indication of the raw materials used or the method of forming.

Most of the early wheels were of the puddled variety (a casting process), as opposed to pressing, and it is doubtful that lead or its compounds were used in this early process.

The grinding wheel industry is no different than most other industries in the dissemination of technical knowledge related to the formulation of bonds and the ingredients used. Most of the research has been directed toward the safety of the product in use. Certainly many of the original materials used are either no longer available or have been replaced due to the progress of ceramic technology, such as the use of frits (low-melting glasses). As a case in point, the Albany slip clay, which we know was used, has been replaced by the use of a more available ball clay and an iron-bearing frit on a 50–50 basis.

This industry is very reluctant to publish percentage composition of bonds such as we can find in any good glaze publication or glass text book on basic formulations for Jena or Pyrex glass.*

Obviously lead oxide (litharge) or compounds of lead were originally incorporated to reduce the maturing temperature of the bond. Most grinding wheels are fired at cone 11 (2280°F), at which temperature regular and semifriable aluminum oxide undergoes a color change from brown to blue. This is due to a migration of titanium to the outer surface of the grain and its partial oxidation. This color could be duplicated by the introduction of lead oxide along with a blue color such as cobalt oxide at a temperature of 2050°F. This seems to be a partial explanation for the use of lead as a bonding ingredient. The fact that some bonds with lead as one of the ingredients have a long firing range (as much as 8 cones) may also have been an added bonus. Some people feel that the product which is fired at cone 1–4 does not dissolve any of the exudation from the abrasive grain that could be dissolved at a higher temperature, namely, 2280°F, thus changing the bond composition and bond characteristics, and ultimately wheel characteristics. A lead-bearing bond would eliminate this problem.

---

*Trade name for borosilicate glass. Manufactured by E. I. du Pont de Nemours & Co., Wilmington, DE.

Probably red lead or litharge was originally used, but with the health hazard accompanying its use, lead bisilicate would be the modern substitute material. Lead-bearing frits of low solubility could also be used.

Litharge has also been used for a grinding aid in rubber-bonded grinding wheels, as well as resinoid-bonded products.

The property of low melting temperatures enhanced the grinding process by providing a lubricating action.

Since there is very little information published on the use of lead or its compounds as a bonding ingredient in vitrified grinding wheels, it is hoped that this short article will be helpful to future scientists.

Reference
[1]William Pinkstone; The Abrasive Ages. Sutter House, Lititz, PA, 1975.

11
Lead Sources and Supplies

Around the world, lead is an important item of commerce and is considered essential to a high standard of living. Lead resources are widely scattered throughout the world. Historically, it has been mined in more than 45 countries and on every continent except Antarctica.

Reserves and Resources

World reserves and resources of lead are shown in Table I. The column headed "Reserves" includes lead in measured and indicated ores that have been inventoried and were considered economic at the time of inventory (1980). Resources, on the other hand, include "inferred" reserves and some hypothetical economic resources in known mining districts, as well as some identified subeconomic resources. If speculative and identified submarginal lead deposits were included, then the total of lead resources could amount to as much as 1.4–1.5 billion tons. H. M. Callaway[1] defined the relationship of ore reserve base to economics as follows:

> "The tonnage of reserves in a given mineralized rock mass depends upon the increment of value that can be recovered *in excess of* the cost of mining and processing the ore. This price/cost differential determines the economic cutoff grade and, therefore, *the tonnage of reserves*. Obviously, the differential can be changed, either by a shift in market price of the mining product, or by a change in the cost of production."

As can be seen from Table I, roughly 130 000 000 metric tons of lead reserves are available worldwide, of which 21% is located in the U.S. (Table II). In 1957 world reserves of lead were estimated to be 45 000 000 metric tons. A decade later, in 1967, lead reserves were still estimated at 44 000 000 metric tons. In the meantime, over 30 000 000 tons of lead had been mined and consumed for industrial applications. In the decade from 1969 to 1979, lead markets generally improved and concomitantly mining technology innovations lowered production costs as intensified exploration proved new mineral finds. The result was an increase of nearly 90 000 000 metric tons of lead reserves during the period. It is highly probable that with continued prospecting in the North American plateau regions, as well as in the developing regions of the world, further significant lead reserves will be found.

The U.S. Bureau of Mines has projected cumulative world demand for lead to be about 115 000 000 metric tons between 1980 and the year 2000. Thus, even if no further lead reserves were located in that 20-year span — a very unlikely event — current world reserves would still prove more than adequate.

Table I. World Lead Resources and Reserves

	Reserves ($\times 10^6$ metric tons)	Resources ($\times 10^6$ metric tons)	Total ($\times 10^6$ metric tons)
North America			
United States	27	47	74
Canada	12	19	31
Mexico	5	5	10
Others	1	1	2
Total	45	72	117
South America			
Brazil	2	2	4
Peru	3	4	7
Others	2	2	4
Total	7	8	15
Europe			
Bulgaria	3	3	6
East Germany	4	5	9
Poland	2	3	5
Spain	3	3	6
Sweden	2	1	3
U.S.S.R.	16	17	33
Yugoslavia	3	3	6
Others	6	5	11
Total	39	40	79
Africa			
Algeria	1	2	3
Morocco	1	2	3
South Africa	6	8	14
Others	1	3	4
Total	9	15	24
Asia			
China (Mainland)	3	4	7
Iran	2	2	4
Others	4	4	8
Total	9	10	19
Oceania			
Australia	18	16	34
Total	18	16	34
World Totals	127	161	288

Table II. United States Lead Reserves by State

State	Lead content (metric ton)
Missouri	19 500 000
Washington, Idaho, Montana	4 700 000
Colorado, Utah, Arizona, California	2 500 000
Other states	300 000
Total	27 000 000

The relative abundance of lead in the Earth's crust (the outer 15 km) has been estimated to be 16 ppm. This is much less than aluminum, iron. or silicon, which together are estimated to make up 40.86% by weight of the earth's crust. Fortunately, the crust is far from uniform due to the action of natural phenomena such as weathering, sedimentation, vulcanism, chemical precipitation, and consolidation. Thus, mineral deposits have been found throughout the world in sufficient concentration for economic extraction (see Table III).

The mineral galena (PbS) is by far the most common lead ore constituent. Two other lead minerals are sometimes found in weathered zones, anglesite ($PbSO_4$), the sulfate, and cerussite ($PbCO_3$), the carbonate, but these are rare and of little commercial significance as ores.

Galena is commonly found associated with other base metal sulfide deposits and rarely by itself. Most often, the zinc sulfide mineral, sphalerite (ZnS), is found as an intimate mixture with galena, and both will often contain silver, iron (FeS), and sometimes gold, copper, bismuth, antimony, and a host of other elements and minerals, e.g., fluorspar and barite. The value of the associated metals often exceeds the value of the lead.

By-Products and Coproducts

U.S. production of lead comes chiefly from ores mined primarily for their lead content. Additional lead is derived from ores in which lead and zinc are coproducts, and it is also recovered as a by-product from ores mined for copper, gold, silver, zinc, or fluorine. The complex ores of the Rocky Mountain areas are particularly dependent on the aggregate value of the lead, zinc, silver, and gold content and not the value of any one metal. Appreciable quantities of sulfur as sulfuric acid also are recovered as by-products of lead production. The relationship of the various associated metals and nonmetals to lead production is shown in Table IV. By-product and coproduct association in lead ores and lead–zinc ores in the rest of the world are similar to the U.S. ores.

World Mine Production

As noted previously, lead is mined in over 45 countries around the world, and approximately 3.5×10^6 metric tons are recovered yearly (see Table III). Nevertheless, more than 60% of the lead mined in the last 6 years came from only

Table III. World Mine Production of Lead

Lead content (×1000 metric tons)	1970	1971	1972	1973	1974	1975	1976	1977	1978	1979	1980	1981
Germany, FRG[†]	50.0	50.1	46.2	45.4	42.7	43.0	42.1	40.9	32.2	33.0	31.3	29
France	28.8	29.8	26.6	25.0	23.5	21.7	28.1	31.5	32.5	29.5	28.8	19
Italy	35.2	31.6	33.7	25.9	23.3	29.5	29.3	31.5	30.5	28.1	23.6	21
United Kingdom[‡]	4.1	5.0	4.5	3.3	2.7	3.5	2.5	2.4	1.8	2.4	2.2	2
Irish Republic	62.8	51.6	59.6	56.2	37.7	36.2	32.6	41.0	47.8	71.1	57.9	29
Norway	3.3	3.2	3.2	3.3	3.4	3.3	3.9	3.2	3.1	3.0	2.6	3
Finland	5.0	4.7	3.8	2.1	1.5	0.9	1.1	0.6	0.8	1.4	1.1	2
Greece	9.2	12.2	16.3	18.8	22.6	14.1	28.2	16.4	22.6	21.7	22.2	23
Yugoslavia	126.7	124.3	120.2	119.3	119.8	126.9	122.5	130.0	124.4	129.8	121.4	119
Austria	5.1	6.2	5.5	5.0	5.8	5.0	4.4	4.3	5.5	5.2	5.5	4
Portugal	1.6	1.4	1.2	0.5								
Sweden	78.3	79.5	75.8	75.8	73.7	70.4	81.6	88.1	81.9	79.4	72.2	85
Spain	72.7	70.2	69.1	63.9	64.1	57.5	66.6	65.3	72.3	74.5	88.6	84
Europe[¶]	482.8	469.8	465.7	444.5[§]	420.8[§]	412.0[§]	442.9[§]	455.2[§]	455.4[§]	479.1[§]	457.4[§]	420
Burma[‡‡]	7.6	8.9	16.3	17.6	11.1	9.8	7.1	8.0	9.2	14.5	14.0	4
India	2.8	3.1	3.7	7.3	7.9	12.3	12.8	13.8	13.1	14.4	14.3	15
Iran	22.9	24.0	33.0	37.5	47.5	50.0	35.0	30.0*	30.0*	15.0*	12.0	20
Japan	64.4	70.6	63.5	52.9	44.2	50.6	51.7	54.8	56.5	46.9	44.8	47
Rep. of Korea	13.4	13.2	11.7	12.9	10.6	12.2	14.5	16.4	15.2	12.4	10.6	12
Thailand	1.3	2.3	1.8	3.7	1.6	1.3	0.8	1.7	3.9	9.0	12.2	17
Turkey	5.8	7.0	7.1	9.6	8.0	6.5	9.5	12.7	10.3	8.5	6.7	8
Other Asia[¶]	1.5	2.5	2.0	4.0	3.1	4.9	5.5	5.4	3.8	1.9	1.8	1
Asia	119.7	131.6	139.1	145.5	134.0	147.6	136.9	142.8	142.0	122.6	116.4	124
Algeria	6.5	4.7	4.8	3.9	3.1	3.2	2.1	0.8	1.5	2.2	2.4	4

Table III. Continued

Lead content (×1000 metric tons)	1970	1971	1972	1973	1974	1975	1976	1977	1978	1979	1980	1981
Kenya*					0.4	1.1	0.5	0.5	0.5			
Congo	0.1		0.5	1.4	1.7	2.0	2.1	2.1	4.8	7.0*	7.0*	6
Morocco	76.1	78.0	86.6	93.2	86.3	69.9	66.5	104.8	109.5	110.7	115.5	118
Namibia	70.5	73.2	59.0	63.3	57.2	53.1	46.4	41.2	38.6	41.0	47.7	47
Nigeria		0.2	0.3	0.4	0.2	0.1	0.1*	0.1*	0.1*			
Rep. of South Africa				1.7	3.3	3.6					85.1	100
Zambia††	27.6	28.3	26.6	25.0	24.6	19.0	14.5	13.4	12.9	12.8	10.0	16
Tunisia	22.5	18.9	18.3	14.8	12.5	10.8	10.5	10.2	8.0	8.8	8.8	6
Africa	203.3	203.3	196.1	203.7	189.3	162.8	142.7	173.1	175.9	182.5	276.5	297
United States‡	518.7	524.9	561.5	547.1	602.3	563.8	553.0	537.5	529.7	525.6	549.5	446
Argentina	35.6	39.9	37.3	35.1	37.8	30.0	33.0	33.6	30.3	31.8	31.6	33
Bolivia**	25.8	23.3	19.2	24.2	19.5	18.0	19.2	18.9	18.0	16.0	17.2	17
Brazil	20.3	27.0	29.9	26.9	25.9	22.4	22.6	24.0	21.2	27.9	25.1	22
Chile	0.9	0.9	0.5	0.3	0.4	0.3	1.8	0.9	0.4	0.3	0.1	0
Greenland‡				5.7	24.1	24.3	27.0	28.8	30.6	31.9	30.1	27
Guatemala	0.1		0.1	0.1	0.1	0.2	0.2	0.1	0.1	0.1	0.1	0
Honduras	15.1	17.4	19.8	20.8	21.5	22.7	21.1	20.6	21.8	16.4	13.3	13
Canada	394.8	393.2	370.0	387.8	301.4	352.5	244.0	327.6	365.8	341.8	296.6	332
Mexico	176.6	156.9	161.4	179.3	218.0	178.6	200.0	163.5	170.5	173.5	145.5	150
Nicaragua		0.5	2.0	2.7	3.2	1.4	1.3	1.0	0.4			
Peru	156.8	165.8	184.4	183.4	165.8	154.2	155.3	171.6	182.7	184.0	189.1	187
Other America	0.3	0.3	0.2	0.2	0.2	0.2	0.2*	0.3	0.2*	0.1	0.2	0
America	1 345.0	1 350.1	1 386.3	1 413.6	1 420.2	1 368.6	1 278.7	1 328.4	1 371.7	1 349.4	1 298.4	1 227

Table III. Continued

Lead content (×1000 metric tons)	1970	1971	1972	1973	1974	1975	1976	1977	1978	1979	1980	1981
Australia‡	456 7	403.6	396.0	402.8	375.3	407.8	397.4	432.2	400.3	421.6	397.9	380
New Zealand	1 0	2.0	1.6	0.6								
Australia and Oceania	457.7	405.6	397.6	403.4	375.3	407.8	397.4	432.2	400.3	421.6	397.9	380
Western countries	2 608.5	2 560.4	2 584.8	2 610.7	2 539.6	2 498.8	2 398.6	2 531.7	2 545.3	2 545.3	2 555.2	2 448
USSR*	470.0	500.0	530.0	570.0	570.0	600.0	600.0	625.0	600.0	590.0	580.0	N.A.
Bulgaria	98.5	100.0*	102.0*	105.0*	110.0	108.0	110.0	110.0	110.0	112.0	110.0	N.A.
Poland	57.0	63.0	73.0	68.0	73.0	77.0	72.0	66.0	63.0	57.0	47.5	N.A.
Romania*	40.0	40.0	45.0	45.0	45.0	34.0	36.0	35.0	33.5	33.5	30.0*	N.A.
Czechoslovakia	6.9	5.8	5.6	5.3	4.4	4.7	4.7	4.3	4.7	3.4	3.0*	N.A.
Hungary	1.8	1.3	1.2	1.4	1.2	0.9	0.9	1.2	1.0	1.0	1.1	N.A.
China, PR*	110.0	120.0	125.0	130.0	130.0	140.0	140.0	150.0	150.0	155.0	160.0	N.A.
North Korea*	70.0	80.0	90.0	100.0	120.0	120.0	110.0	110.0	110.0	120.0	125.0	N.A.
Eastern countries	854.2	910.1	971.8	1 024.7	1 053.6	1 084.6	1 073.6	1 101.5	1 072.2	1 071.9	1 056.6	1 002*
Total World	3 462.7	3 470.5	3 556.6	3 635.4	3 593.2	3 583.4	3 472.2	3 633.2	3 617.5	3 627.1	3 603.2	3 450

Source: "Metal Statistics." 61st Edition, published by Metallgesellschaft A.G.

*Estimates.
‡Excluding lead content in pyrites.
‡Recoverable metal content.
§Excluding Greenland.
¶Excluding "Eastern countries".
**Up to 1972, exports; since 1973, production.
††Smelter production.
‡‡Since 1979 mine production.

138

Table IV.   U.S. Lead By-Product and Coproduct Relationship in 1978

Source	Product	Unit	Quantity	Percent of total output
Lead	Bismuth	Metric tons	W*	100.0
Lead	Antimony	Short tons	539	40.3
Lead	Zinc	Metric tons	75 238	25.0
Lead	Silver	Troy ounces	3 420 374	8.7
Lead	Tellurium	Metric tons	W*	W*
Lead	Copper	Metric tons	11 040	0.8
Lead	Sulfur	Metric tons	66 969	0.6
Lead	Gold	Troy ounces	682	0.1
Lead	Lead	Metric tons	478 268	90.3
Zinc	Lead	Metric tons	32 221	6.1
Silver	Lead	Metric tons	18 308	3.4
Copper	Lead	Metric tons	325	†
Fluorine	Lead	Metric tons	363	†
Gold	Lead	Metric tons	176	†
Tungsten	Lead	Metric tons	W*	W*

*W = Withheld to avoid disclosing company proprietary data.
†Less than ½%.

6 countries. Of these, U.S.S.R. and the United States produced more than 30% (see Table V). It is interesting to note that in the period 1925–1929, world lead mine production averaged 1 635 000 metric tons, and only four countries, the United States with 37%, Mexico with 14%, Australia with 12%, and Canada with 8.5%, mined more than 71% of the world total.

Lead and lead–zinc mines in the above-mentioned countries, except for a few scattered mining operations around the world, are all underground. A summary of U.S. mine production by state of origin for the years 1970–81 is given in Table VI.

World Refined Lead Production

Once the lead is mined, the ore is crushed, ground, and prepared for flotation, the method which is generally used to separate the lead sulfide from the rock matrix and other minerals and to concentrate it to about 60–70%.

The concentrated material serves as feed for the lead smelter, which treatment includes sintering, smelting, drossing, and refining. The sintering step agglomerates the fine concentrates and removes most of the sulfur as sulfur dioxide, which at most smelters goes to an acid plant to be converted to sulfuric acid.

The sintered product, along with coke, fluxes, and dross, is smelted in a blast furnace to produce impure lead bullion, slag, and fume. The lead goes from the blast furnace to the drossing kettle. The slag is crushed and treated in a slag fuming plant to recover zinc oxide fume. A drossing kettle and furnace are used to reduce

Table V. Mine Production of Lead

Country	Lead content ($\times 10^3$ metric tons)						% Total (6-year av.)
	1975	1976	1977	1978	1979	1980	
USSR	600	600	625	600	590	580	16.7
United States	564	553	538	530	526	550	15.2
Australia	408	397	432	400	422	398	11.4
Canada	353	244	328	366	342	297	9.0
Peru	154	155	172	183	184	189	4.8
Mexico	179	200	164	171	174	146	4.8
Ratio = $\dfrac{\text{6 Leaders}}{\text{Total}}$	$\dfrac{2258}{3583}$ =63%	$\dfrac{2149}{3472}$ =61.9%	$\dfrac{2259}{3633}$ =62.2%	$\dfrac{2250}{3618}$ =62.2%	$\dfrac{2238}{3627}$ =61.7%	$\dfrac{2160}{3603}$ =60%	$\dfrac{2220}{3589}$ =61.9%

Table VI. United States Mine Production of Lead by States (Lead Content in Metric Tons)

	1970	1971	1972	1973	1974	1975	1976	1977	1978	1979	1980	1981
Arizona	258	779	1 599	692	961	381	307	288	416	354	401	736
California	1 607	2 072	1 046	40	32	60	49	3		2		
Colorado	19 827	23 356	28 437	25 503	22 325	24 574	24 266	20 860	15 151	7 554	10 272	11 339
Idaho	55 530	60 428	55 708	56 013	46 917	45 718	48 658	42 872	44 761	42 636	38 607	38 496
Illinois	1 390	1 123	1 211	490	447							
Kansas	73											
Maine			77	185	253	330	196	161				
Missouri	382 618	389 758	443 974	441 929	509 926	468 069	454 491	453 824	461 762	472 054	497 170	388 541
Montana	904	558	260	160	140	186	83	96	132	258	295	190
Nevada	330	101			1 619	2 700	528	674	653	24	26	
New Mexico	3 221	2 695	3 249	2 318	2 145	1 752				43		
New York	1 161	795	988	2 090	2 791	2 746	2 899	2 520	990	458	876	
Oklahoma	723											
Utah	41 165	34 718	18 784	12 458	9 535	11 502	14 790	9 749	2 541			
Virginia	3 044	3 072	3 122	2 392	2 818	2 314	1 765	1 999	1 803	1 596	1 563	
Washington	6 154	4 696	2 328	2 011	1 178			1 090				
Wisconsin	690	682	687	767	1 166	736						
Other states	3	19		5	2	2 715	4 943	3 364	1 452	590	274	4 773
Total	518 698	524 852	561 470	547 053	602 255	563 783	552 975	537 500	529 661	525 569	549 484	444 075

Source: U.S. Bureau of Mines.

impurities and to remove copper from the bullion by controlled cooling to the point where copper becomes insoluble. The dross, containing copper, is treated in a reverberatory furnace to make a copper matte, and the lead bullion is ready for refining.

Refining removes silver and any remaining copper through the addition of zinc dust. The last traces of zinc are removed by vacuum dezincing; caustic is added to remove minor amounts of arsenic and antimony. Bismuth is removed from lead bullion, when required, by the Betterton–Kroll process through the addition of calcium and magnesium. The refined lead product contains more than 99.9% lead.

An alternate electrolytic refining process, called the Betts process, is used by smelters in Canada and Peru. In this process, anodes of lead bullion are electrolyzed in a solution of lead fluosilicate and free fluosilicic acid, and the electrodeposited lead on the cathode is melted, drossed, and cast into refined lead pigs.

The Imperial smelting process was developed in England to smelt sintered mixed lead–zinc concentrates from complex lead–zinc ores. This blast furnace technique produces zinc metal, lead bullion containing gold and silver, and matte. Feed to the furnace can include oxidized ores and secondary materials as well as various mixes of lead and zinc concentrates. Furnaces of this type are now operating in 11 countries.

The average yearly world smelter production in the period 1925–29 was 1.64×10^6 metric tons. In 1933 smelter output dropped to 1.15×10^6 metric tons owing to the abnormal economic situation. Under the stimulus of World War II, it rose to 1.74×10^6 metric tons in 1940 before the German offensive destroyed productive will and ability in many European smelting areas (see Table IX).

The first postwar year, 1946, saw smelter output fall to 1.16×10^6 tons — the lowest world output since 1922. In that year strikes, unsure price structure, cancellation of war-generated contracts, and other marks of industrial transition to a peacetime economy adversely affected smelter output. Reconstruction needs for raw materials in many parts of the world and long-delayed needs of the consumer-goods industries in the United States caused large yearly increases in world smelter output during the years 1947–50. Abnormally large demands for lead following the outbreak of the Korean war stimulated smelter output to records in 1952 and 1953. The years following have seen continued expansions. In 1960 total world smelter output was 2.72×10^6 metric tons. By 1971 production of refined lead had broken a new mark, reaching nearly 4×10^6 metric tons, and 5 years later, in 1976, refined lead totaled 5×10^6 metric tons, an increase of nearly 3×10^6 metric tons in 30 years! The year 1979 saw record production of 5.65×10^6 tons, and that mark is not expected to be reached again for a decade.

As with the mining of lead, refining is also widespread, owing partly to the ease with which it is reclaimed and to the costs associated with shipping the heavy metal long distances.

As shown in Table VII, nearly 60 countries reported lead refining operations in the period 1975–1981. Nevertheless, as with mine production, only a few countries account for more than 50% of production. As it has for many years, the U.S. leads in world production of refined lead, supplying about 21% of the total

Table VII. Production of Refined Lead (Including Secondary Antimonial Lead)
(Metric Tons ×1000)

Country	1975[¶]	1976	1977	1978	1979	1980	1981
Germany, FR	316.3	337.4	373.5	369.0	373.3	350.3	348
Belgium-Luxembourg	103.0	105.7	106.0	104.0	92.2	105.9	102
Denmark	13.7	14.6	24.2	26.2	29.8	24.5	27
France	180.3	205.2	217.8	222.8	219.7	218.8	228
United Kingdom	313.4	341.9	351.1	345.8	368.3	324.8	333
Italy	90.0	118.2	117.7	116.2	126.2	133.7	133
Irish Rep.*	5.0	5.0	5.0	5.0	7.0	7.0	10
Netherlands	37.0	36.7	33.8	31.9	30.1	27.7	20
Finland*	4.0	5.0	5.0	5.0	6.0	3.2	7
Greece	16.3	18.7	20.4	22.6	22.0	21.1	21
Yugoslavia	126.1	111.2	129.9	116.7	111.0	101.8	126
Norway	0.4	0.6	0.5	0.3	0.4	0.3	0
Austria	16.0	18.4	19.1	17.6	19.9	17.9	17
Portugal	1.6	0.6	0.4	0.3	1.0	1.0	5
Spain	100.9	101.7	118.6	122.2	127.0	120.7	120
Sweden	37.1	33.2	42.6	45.3	46.7	42.3	29
Switzerland*	5.0	5.0	5.0	5.0	5.0	7.0	7
Europe[†]	1366.1	1459.1	1570.6	1555.9	1585.6	1508.0	1533
Burma	2.4	2.7	5.2	4.8	6.1	5.9	4
Cyprus*	2.5	2.5	2.5	2.5	2.5		
India	15.0	15.0	20.0	20.6	29.0	35.6	25
Indonesia	1.0	1.0	2.0	2.0	2.0	2.0	4
Japan	252.4	281.5	287.7	291.1	282.7	304.9	317
Republic of Korea	5.7	7.9	7.0	9.0	13.4	12.0	21
Malaysia*	2.0	2.0	2.0	2.0	2.0	2.3	
Pakistan*	1.5	1.5	1.5	1.5	1.5	1.0	
Philippines	3.1	3.3	3.4	3.5	3.6	4.8	4
Sri Lanka*	1.0	1.0	1.0	1.0	1.0	1.0	
Taiwan	5.0	8.0	10.0	11.0	20.0	16.8	30
Thailand	0.9	0.8	1.2	1.1	1.5	1.5	5
Turkey	3.0	3.2	3.0	3.0	5.9	6.5	5
Other	0.3	0.3	0.3				11
Asia[†]	295.8	330.7	346.8	353.1	371.2	394.3	422
Morocco	9.2	26.3	34.6	30.0	36.8	42.4	57[3]
Namibia	44.3	39.6	42.7	39.5	41.7	42.7	
Nigeria*					1.5	2.0	2
Rep. of South Africa	13.2	22.4	24.3	23.6	23.3	35.4	67[‡]

143

Table VII. Continued

Country	1975[¶]	1976	1977	1978	1979	1980	1981
Zambia	19.0	14.5	13.4	12.9	12.8	10.0	10
Tunisia	23.7	24.2	19.7	16.6	17.7	19.2	18
Africa	109.4	127.0	134.7	122.6	133.8	151.7	154
United States	1056.8	1107.2	1169.1	1188.4	1225.6	1150.5	1067
Argentina	46.7	50.0	45.0	29.7	56.0	46.7	35
Bolivia	0.1		0.1	0.5	0.5*	1.0*	§
Brazil	62.7	69.2	77.3	80.4	98.1	85.0	66
Jamaica	1.5	1.5	2.0	2.0	2.0	1.0	§
Canada	216.3	231.0	240.6	245.9	252.4	234.6	238
Colombia*	1.5	1.5	1.5	2.0	2.5	3.0	§
Mexico	164.0	173.0	206.0	226.0	224.6	184.7	166
Peru	76.0	79.1	84.3	79.6	90.7	87.3	85
Trinidad*	1.5	1.5	1.5	2.0	2.0	2.0	§
Venezuela*	4.0	5.0	6.0	8.0	10.0	10.0	10
Other							9
America	1631.1	1719.0	1833.4	1864.5	1964.4	1805.8	1676
Australia	193.5	211.5	218.0	234.1	257.7	233.1	245
New Zealand*	12.0	12.0	12.0	12.0	12.0	12.0	7
Australia and Oceania	205.5	223.5	230.0	246.1	269.7	245.1	252
Western countries	3607.9	3859.3	4115.5	4142.2	4324.7	4104.9	4037
USSR*	700.0	700.0	720.0	770.0	780.0	780.0	
Bulgaria*	110.0	112.0	120.0	125.0	120.0	118.0	
Germany, DR*	39.0	42.0	45.0	45.0	42.0	42.0	↑
Poland	76.2	80.6	85.4	86.7	82.0	85.0	Not
Romania	38.9	42.4	41.7	42.8	40.9	42.0*	Avail-
Hungary	0.7	3.0	1.6		0.1	0.1	able
Czechoslovakia	18.4	19.1	19.0	19.0	19.0	18.0*	↓
China, PR*	140.0	140.0	150.0	160.0	170.0	175.0	
North Korea*	80.0	70.0	70.0	75.0	70.0	65.0	
Eastern countries	1203.2	1209.1	1252.7	1323.5	1324.0	1325.1	1300*
World total	4811.1	5068.4	5368.2	5465.7	5648.7	5430.0	5337*

*Estimate.
[†]Excluding "Eastern countries".
[‡]Includes Namibia.
[§]Included in "Other."
[¶]World production (in metric tons ×1000) for 1970–1974 was 3988, 3937, 4086, 4224, and 4281, respectively.

Table VIII. World Leaders' Production (metric tons, $\times 10^3$)
of Refined Lead (Including Secondary)

Country	1975	1976	1977	1978	1979	1980	% Total
United States	1057	1107	1169	1188	1226	1151	21.2
USSR	700	700	720	770	780	780	14.4
Germany, FR	316	337	374	369	373	350	6.4
United Kingdom	313	342	351	346	368	325	6.0
Japan	252	282	288	291	283	305	5.6
Total	2638	2768	2902	2982	3030	2911	53.6
All Others	2173	2300	2466	2484	2619	2519	
Grand total	4811	5068	5368	5466	5649	5430	

in 1980. The U.S.S.R. followed with 14%, trailed by Germany, U.K., and Japan with about 6% each (see Table VIII).

World Consumption

Statistics on world consumption of lead (See Tables IX and X) are less accurate than refinery production data due to many factors, e.g., lack of uniformity in reporting methods, estimation by some countries, and weak interest in the process by consuming plants in some parts of the world. These data, therefore, should only be relied upon for year-to-year comparisons, not as absolutes.

For the 40 years preceding 1971, Table IX shows the United States to be the leading consumer of refined lead, averaging between 30 and 40% of world total. Since 1972, the percentage appears to have dropped. This is due in part to use of a new reporting basis showing greater consumption worldwide. U.S. consumption as a percent of world total is shown in Table XI.

United States lead consumption by industrial use for the 55-year period, 1925–1980, is shown in Table XII. Major uses for lead have changed considerably over this period. One notes that 1925 showed white lead at 118 000 metric tons and cable sheathing at 142 000 metric tons, whereas in 1981 these two markets together did not equal 20 000 metric tons. In the meantime, a market that did not show on the U.S. Bureau of Mines consumption table until 1941, lead antiknock (tetraethyllead), grew to reach a high of 252 700 metric tons in 1970. Another growth market worldwide was the storage battery. Consumption figures for 1925 showed 163 000 metric tons and in 1981 batteries had grown to 770 154 metric tons (Table XIII). Consumption of lead in the U.S. by industries is given in Table XIV. Other significant changes occurred in automobile-related markets due in part to substitution, weight reduction efforts by the auto industry, and a shift in location of manufacture, viz., from the U.S. to Japan and Europe. The reduction from 9.3×10^6 cars to less than 6×10^6 in 1982 produced significant dislocations in the secondary scrap market as well as reduced the number of new batteries required for OEM manufacturing.

Secondary Lead Resources

Although modest amounts of lead have no doubt been reclaimed from scrap since 3000 B.C., it was not until the advent of the lead acid battery and its adoption

Table IX. World Production and Consumption of Lead for 1900–1980
(in Metric Tons ×1000)

Year	Mine* prodctn	Prodctn of refined lead	Consmptn	Year	Mine* prodctn	Prodctn of refined lead	Consmptn
1900		871		1941	1664	1724	1700
1901		861		1942	1642	1664	1664
1902		882		1943	1479	1482	1492
1903		903		1944	1322	1287	1286
1904		970	981 av.	1945	1175	1180	1136
1905		969		1946	1144	1154	1297
1906		984		1947	1351	1394	1526
1907		1015		1948	1440	1490	1503
1908		1055		1949	1540	1665	1399
1909		1087		1950	1686	1850	1866
1910		1127		1951	1733	1878	1772
1911	1198	1132		1952	1874	2030	1743
1912	1212	1218		1953	1969	2186	1894
1913	1223	1200		1954	2079	2344	2185
1914	1096	1183		1955	2178	2411	2370
1915	1100	1150	1125 av.	1956	2246	2560	2427
1916	1104	1154		1957	2336	2629	2426
1917	1149	1185		1958	2309	2618	2359
1918	1167	1223		1959	2302	2563	2552
1919	857	869		1960	2376	2717	2616
1920	859	873		1961	2423	2822	2694
1921	818	858	930	1962	2530	2776	2789
1922	1055	1045	1022	1963	2547	2949	2923
1923	1186	1176	1180	1964	2570	3100	3148
1924	1323	1299	1365	1965	2741	3208	3183
1925	1501	1510	1446	1966	2901	3318	3331
1926	1579	1603	1510	1967	2903	3373	3328
1927	1649	1673	1618	1968	3034	3545	3675
1928	1609	1660	1702	1969	3277	3860	3843
1929	1688	1763	1726	1970	3463	3988	3924
1930	1603	1659	1601	1971	3471	3937	4002
1931	1336	1369	1317	1972	3557	4086	4139
1932	1191	1152	1094	1973	3635	4224	4387
1933	1163	1150	1185	1974	3593	4281	4420
1934	1299	1324	1375	1975	3583	4811[†]	4718.8[†]
1935	1369	1380	1430	1976	3472	5068	5138.3
1936	1482	1475	1584	1977	3633	5368	5443.9
1937	1674	1697	1681	1978	3618	5466	5517.0
1938	1779	1642	1699	1979	3627	5649	5607.3
1939	1718	1717	1731	1980	3603	5430	5311.0
1940	1762	1736	1756				

*Metal content.

[†]Up to 1975 new reporting basis.

Sources: "Metal Statistics," various editions—Published by Metallgesellschaft A.G. and the American Bureau of Metal Statistics, U.S. Dept. of Interior, Bureau of Mines—IC 8083.

Table X. Consumption of Refined Lead (in Metric Tons ×1000)

Country	1975	1976	1977	1978	1979	1980
Germany, FR	282.5	299.8	348.5	355.8	361.3	333.1
Belgium-Luxembourg	49.7	55.1	59.0	56.1	57.6	59.3
France	190.3	228.1	210.4	211.7	211.4	212.0
Italy	192.0	265.0	260.0	251.0	258.0	275.0
Netherlands	44.4	45.7	56.8	55.9	54.7	58.0
Denmark	18.6	17.1	21.9	19.0	25.0	25.0
United Kingdom[‡]	306.0	318.3	317.7	336.5	333.2	295.5
Irish Republic	6.7	7.3	7.4	6.0	5.8	8.6
Iceland	0.1	0.1	0.1	0.4	0.3	0.2
Norway	14.8	15.0	14.5	14.5	13.0	13.5
Finland	17.9	19.1	19.4	16.5	20.8	22.8
Greece	22.8	27.8	28.5	28.6	26.4	27.6
Yugoslavia	83.0	77.0	88.0	82.0	84.0	84.6
Austria	31.5	41.2	41.5	42.4	46.6	52.9
Portugal	9.8	15.3	12.9	16.5	12.0	16.0
Sweden	29.5	21.8	24.5	18.3	22.0	20.5
Switzerland	17.6	18.2	16.1	17.3	18.5	20.9
Spain	95.2	115.0	122.0	114.8	115.0	110.6
Europe[†]	1412.4	1586.9	1649.2	1623.3	1665.6	1636.1
India	36.0	52.2	54.0	55.0	58.4	56.0
Iran*	14.4	11.4	12.0	11.0	10.0	2.2
Japan	260.7	309.7	333.9	352.1	368.2	392.5
Rep. of Korea	10.3	10.6	17.8	27.4	33.1	30.0
Philippines	9.5	8.3	9.5	11.0	10.8	10.0*
Taiwan	9.5	13.8	17.3	19.8	27.0*	16.1
Thailand	8.0	9.3	7.0	7.0	8.9	12.8
Turkey	7.7	10.7	10.4	7.0	6.7	11.0
Other Asia[†]	25.9	26.4	37.3	32.6	38.2	44.6
Asia[†]	382.0	452.4	499.2	522.9	561.3	575.2
Algeria	4.5	4.5	4.1	9.5	5.1	6.4
Rep. of South Africa	40.7	34.8	36.7	39.6	43.8	52.6
Tunisia	4.6	5.2	4.6	5.2	5.6	5.1
Other Africa	26.7	19.8	26.4	29.1	30.0*	28.0
Africa	76.5	64.3	71.8	83.4	84.5	92.1
United States	1122.7	1272.3	1417.9	1405.5	1345.4	1094.0
Argentina	52.8	46.0	45.0	38.2	33.6	27.5
Brazil	75.2	79.4	92.1	81.2	96.8	82.7
Canada	89.2	107.6	107.0	100.8	122.0	104.4
Mexico	72.3	81.0	88.0	108.0	110.1	96.2
Peru	12.7	13.3	13.2	14.5	24.0	26.1

Table X. Continued

Country	1975	1976	1977	1978	1979	1980
Other America	18.6	16.4	18.2	24.8	27.5	26.5
America	1443.5	1616.0	1781.4	1773.0	1759.4	1457.4
Australia	72.3	75.2	76.8	72.0	78.7	70.5
New Zealand	19.6	20.5	19.8	19.6	18.6	18.3
Other Oceania	1.0	0.5		1.0		
Australia and Oceania	92.9	96.2	96.6	92.6	97.3	88.8
Western countries	3407.3	3815.8	4098.2	4095.2	4168.1	3849.6
USSR*	700.0	700.0	720.0	760.0	780.0	800.0
Albania*	2.0	2.5	2.5	2.5	2.5	2.5
Bulgaria	95.0	95.0	95.0	105.0	108.0	108.0
Germany, DR*	91.0	93.0	95.0	98.0	100.0	100.0
Poland	95.0	92.0	89.0	96.2	86.7	88.6
Romania*	48.0	48.0	45.0	45.0	47.0	50.0
Czechoslovakia	53.6	55.2	54.2	57.5	59.0	58.0
Hungary	13.9	15.5	13.6	13.9	12.0	12.3
China, PR*	185.0	190.0	200.0	210.0	210.0	210.0
North Korea*	25.0	28.0	28.0	30.0	30.0	28.0
Cuba	2.2	2.5	2.5	2.7	3.0	3.0
Other eastern countries	0.8	0.8	0.9	1.0	1.0	1.0
Eastern countries	1311.5	1322.5	1345.7	1421.8	1439.2	1461.4
Total World	4718.8	5138.3	5443.9	5517.0	5607.3	5311.0

*Estimated.
†Excluding "Eastern countries".
‡Excluding scrap and remelted lead.

Source: "Metal Statistics" by Metallgesellschaft A.G.

as the power source for electric vehicles (1908–1915) that a huge resource base for recycling lead began to develop. Indeed, the secondary lead industry has risen from insignificance in 1900 to the point where in 1980 more than 2 million tons of lead were recycled worldwide. In the U.S., 673 000 tons of lead were reclaimed in 1980, representing about 62% of consumption.

Shortly after electric vehicles had reached their primacy, Charles "Boss" Kettering invented the battery/electric starter for gasoline vehicles, ushering in a huge *new* market for lead — starting, lighting, ignition (SLI) batteries in gasoline engine powered automobiles.

Table XI. United States Lead Statistics, 1931–80 (Metric Tons Unless Otherwise Specified)

Year	Mine production	Refinery production (primary sources)	Secondary production	General imports (refined)*	Exports	Reported consumption	Price (cents per pound)†	U.S. consumption as % of world total
1931	367 067	401 669	212 916	9	19 654	515 009	4.24	39
1932	265 776	255 773	179 895	40	21 333	378 024	3.18	
1933	247 368	239 203	203 663	99	20 716	407 780	3.87	
1934	260 670	282 349	189 057	257	5 361	442 706	3.86	
1935	300 372	294 436	245 303	1 199	6 334	488 882	4.06	
1936	338 307	362 108	238 499	2 350	16 613	574 792	4.71	
1937	421 743	423 943	249 567	4 448	18 226	615 706	6.01	
1938	335 410	348 059	204 025	2 935	41 609	495 323	4.74	
1939	375 556	439 109	219 085	6 476	67 487	605 092	5.05	
1940	414 939	483 692	236 182	137 482	21 550	709 419	5.18	40
1941	418 599	517 973	360 530	248 740	13 026	952 544	5.79	
1942	450 181	514 228	293 022	332 481	1 760	946 194	6.48	
1943	411 239	426 025	310 343	221 816	1 817	1 009 697	6.50	
1944	378 170	421 626	300 656	202 083	14 082	1 014 777	6.50	
1945	354 556	402 414	329 344	206 356	1 277	953 996	6.50	
1946	304 338	306 807	356 330	104 783	542	867 722	8.11	
1947	348 560	400 078	464 452	144 708	1 382	1 063 221	14.67	
1948	354 234	368 947	453 657	224 180	362	1 028 653	18.04	

149

Table XI. Continued

Year	Mine production	Refinery production (primary sources)	Secondary production	General imports (refined)*	Exports	Reported consumption	Price (cents per pound)†	U.S. consumption as % of world total
1949	371 862	433 034	373 926	249 694	879	868 787	15.36	
1950	390 840	451 135	437 513	400 783	2 481	1 123 078	13.30	
1951	352 137	378 925	470 022	162 415	1 162	1 074 826	17.49	60
1952	353 949	428 964	427 551	463 318	1 598	1 025 840	16.47	
1953	310 841	424 464	441 561	349 331	728	1 090 077	13.48	
1954	295 215	441 538	436 288	250 643	541	993 251	14.05	
1955	306 651	434 684	455 453	239 632	366	1 100 092	15.14	
1956	320 078	431 974	459 721	238 276	4 198	1 097 437	16.01	
1957	306 824	434 013	443 821	294 181	3 936	1 032 481	14.66	
1958	242 560	426 518	364 495	334 254	1 233	894 836	12.11	
1959	231 864	309 287	409 492	238 967	2 500	989 874	12.21	
1960	223 774	346 940	426 289	186 910	1 784	926 392	11.95	
1961	237 611	437 839	410 766	233 012	1 935	931 875	10.87	35
1962	214 963	341 159	402 973	233 329	1 912	1 006 644	9.63	
1963	229 853	358 095	447 669	205 955	987	1 055 381	11.14	
1964	259 464	437 715	491 315	188 553	9 231	1 090 562	13.62	
1965	273 196	379 429	522 374	201 951	7 086	1 126 254	16.00	
1966	296 983	399 828	519 666	258 901	4 931	1 201 001	15.12	

Table XI. Continued

Year	Mine production	Refinery production (primary sources)	Secondary production	General imports (refined)*	Exports	Reported consumption	Price (cents per pound)†	U.S. consumption as % of world total
1967	287 515	344 634	502 374	329 851	5 929	1 143 521	14.00	
1968	325 821	423 937	499 749	306 737	7 512	1 205 458	13.21	
1969	461 769	579 378	547 854	252 542	4 507	1 260 405	14.93	
1970	518 698	604 847	541 943	221 918	7 028	1 234 272	15.69	
1971	524 852	589 684	541 405	177 434	5 375	1 298 648	13.89	32
1972	561 470	617 248	559 368	219 893	7 599	1 347 400	15.03	
1973	547 054	611 911	593 558	161 585	60 397	1 398 162	16.29	32
1974	602 253	610 557	633 848	107 380	56 229	1 450 976	22.53	
1975	563 783	577 080	597 341	91 182	19 283	1 176 708	21.53	25
1976	552 971	592 280	659 132	132 387	5 332	1 351 771	23.10	
1977	537 499	548 700	757 592	237 023	8 931	1 435 473	30.70	26
1978	529 661	565 173	769 236	221 313	8 225	1 432 744	33.65	
1979	525 569	575 611	801 368	191 662	10 646	1 358 335	52.64	24
1980	549 484	549 707	673 458	81 733	164 458	1 070 303	42.46	20

*1931–39 includes a small quantity of scrap.
†Quotations for 1931–71 at New York and from 1982 on a nationwide, delivered basis.

Source: U.S. Bureau of Mines.

Table XII. Lead Consumption in the United States by Uses, 1925–59 (in Short Tons ×1000)*

Use	1925	1926	1927	1928	1929	1930	1931	1932	1933	1934	1935	1936
Ammunition	31	32	34	40	41	33	30	23	32	35	29	32
Automobiles	13	17	12	17	18	11	6	3	5	7	10	11
Bearing metal	34	36	31	32	33	20	12	10	11	12	13	16
Building	83	94	88	96	96	67	40	22	26	30	32	40
Cable covering	155	185	161	180	220	208	117	55	31	35	39	61
Calking	30	32	30	32	31	21	15	10	12	10	12	13
Castings	13	18	17	18	18	12	7	5	5	5	5	6
Foil	32	35	30	35	40	26	20	14	22	16	16	28
Red lead and litharge†	42	36	38	31	30	32	18	32	38	42	47	54
Solder	35	37	35	37	37	27	20	14	16	16	20	22
Storage batteries	180	190	175	220	210	163	157	138	147	163	175	191
Terne plate	4	4	4	4	4	3	2	1	2	3	5	6
Tetraethyl												
Type metal	15	16	16	17	18	16	14	11	11	13	15	17
White lead	131	120	126	123	120	84	77	54	59	64	80	85
Other uses	47	49	44	49	56	46	33	25	33	37	41	52
Total	856	901	841	931	972	769	568	417	450	488	539	634

Use	1937	1938	1939	1940	1941	1942	1943	1944	1945	1946	1947
Ammunition	39	31	42	56	71	83	178	63	43	30	40
Automobiles	12	6	9								

Table XII. Continued

Use	1937	1938	1939	1940	1941	1942	1943	1944	1945	1946	1947
Bearing metal	15	9	13	14	25	20	35	41	42	41	40
Building	45	36	50	65	95	110	62	70	68	78	67
Cable covering	90	60	74	107	173	165	141	131	107	111	159
Calking	15	12	16	19	31	40	30	32	35	39	50
Casting	6	6									
Foil	22	22	22	24	45	8	13	16	16	4	4
Red lead and litharge†	57	43	57	59	89	68	79	80	68	61	66
Solder	22	15	20	24	36	38	38	41	47	53	59
Storage batteries	192	167	198	220	245	216	257	307	295	260	380
Terne plate	6	4	6	6	9	5	5	6	5	3	(5)
Tetraethyl					50	49	60	85	76	48	67
Type metal	17	12	14	17	20	20	17	23	25	33	26
White lead	86	71	75	66	85	75	50	60	41	48	55
Other uses	55	52	71	105	76	146	148	164	184	147	159
Total	679	546	667	782	1050	1043	1113	1119	1052	956	1172

Use	1948	1949	1950	1951	1952	1953	1954	1955	1956	1957	1958	1959
Metal products:												
Ammunition	50	24	39	40	36	45	40	47	45	43	40	46
Bearing metal	43	29	38	35	37	39	27	35	28	27	19	23
Brass and bronze	23	15	22	30	26	26	20	24	27	24	20	24

153

Table XII. Continued

Use	1948	1949	1950	1951	1952	1953	1954	1955	1956	1957	1958	1959
Cable covering	172	144	132	132	143	146	128	121	134	108	75	62
Calking lead	31	35	53	47	45	48	50	59	65	66	71	80
Casting metal	9	13	19	22	18	13	11	15	13	13	9	8
Collapsible tubes	11	9	13	14	10	12	11	11	11	10	8	9
Foil	3	2	4	3	2	4	4	5	5	5	5	4
Pipes, traps, and bends	40	30	41	33	29	29	27	30	28	25	23	25
Sheet lead	32	27	31	31	29	30	26	30	30	27	25	28
Solder	71	62	95	83	73	79	71	89	75	71	60	69
Terne metal	3	3	4	2	2	3	1	2	2	1	1	2
Type metal	26	21	25	28	27	27	26	27	27	29	27	28
Total	514	414	516	500	477	501	442	495	490	449	383	408
Storage batteries:												
Grids, posts, etc.	204	175	212	200	188	192	174	196	192	186	160	187
Oxides	150	139	186	175	163	176	163	184	179	175	153	194
Total	354	314	398	375	351	368	337	380	371	361	313	381
Pigments:												
Pigment colors	11	8	13	13	13	13	14	15	14	12	12	14
Red lead and litharge	80	71	102	88	77	89	76	87	79	78	65	74
White lead	31	18	36	26	23	18	18	19	17	16	13	11

Table XII. Continued

Use	1948	1949	1950	1951	1952	1953	1954	1955	1956	1957	1958	1959
Other[†]	20	10	15	13	9	10	8	10	10	9	6	5
Total	142	107	166	140	122	130	116	131	120	115	96	104
Chemicals:												
Tetraethyl lead	84	95	114	128	147	162	160	165	192	177	159	160
Miscellaneous chemicals	10	4	12	7	4	7	7	6	3	4	4	5
Total	94	99	126	135	151	169	167	171	195	181	163	165
Miscellaneous uses:												
Annealing	6	5	6	7	5	5	5	6	6	5	5	5
Galvanizing	2	1	2	2	2	2	3	2	2	1	1	1
Lead plating	2	1	2	1	1	1	1	1	1	1	1	
Weights and ballasts	6	5	7	8	8	9	7	8	7	8	7	9
Total	16	12	17	18	16	17	16	17	16	15	14	15
Other, unclassified	14	12	15	17	14	17	17	19	18	17	17	18
Grand total	1134	958	1238	1185	1131	1202	1095	1213	1210	1138	986	1091

[1]Source: Bureau of Mines Minerals Yearbooks.
[2]Includes lead content of leaded zinc oxide and other nonspecified pigments.

155

Table XIII. Lead Consumed in U.S. Battery Manufacture

Year	Lead consumption (short tons)	% Total consumption
1970	538 374	43.6
1971	616 710	47.5
1972	659 156	48.9
1973	698 034	49.9
1974	772 817	53.3
1975	634 501	53.9
1976	746 076	55.2
1977	858 089	59.8
1978	879 278	61.4
1979	814 335	60.0
1980	645 360	60.3
1981	770 154	66.0

Source: U.S. Bureau of Mines.

U.S. Secondary Resource Pool

H. M. Calloway, in his notable "LEAD: A Materials Survey,"[1] stated that

"after suitably factoring the 1919–1939 apparent consumption figures and adjusting for probable average service life, approximately 900,000 short tons of recoverable lead from that period was in the secondary resource pool as of 1960.
Summation of the yearly increments 1939 through 1959 indicates that approximately 3 million additional tons of recoverable lead were added to the secondary resources pool in that 20 year period. Thus, the total quantity of lead in use in the U.S., constituting the U.S. Secondary Resource Pool in 1960, was 3,900,000 short tons."

The production of secondary lead in the U.S. is given in Table XV.

Worldwide Secondary Resource Pool

Calloway proceeded from that point to estimate the world secondary resource pool with the following rough formula:

$$\frac{\text{World pool}}{\text{U.S. pool}} = \frac{\text{World consumption}}{\text{U.S. consumption}}$$

Using the 3.9 million ton estimate to 1960, Calloway arrived at the following world pool estimate:

$$\text{World pool} = \frac{\overset{\text{(World consumption)}}{2\,500\,000} \times \overset{\text{(U.S. pool)}}{3\,900\,000}}{\underset{\text{(U.S. consumption)}}{1\,200\,000}} = 8\,100\,000 \text{ tons}$$

Table XIV. U.S. Consumption of Lead by Industries[†]

1000 kilos (1 metric ton)	1970	1971	1972	1973	1974	1975	1976	1977	1978	1979	1980*
Ammunition	65 976	79 439	76 838	73 917	79 007	68 112	66 658	62 042	55 776	53 236	48 662
Bearing metals	14 813	14 773	14 438	14 204	13 253	11 053	11 851	10 873	9 510	9 630	7 808
Brass and bronze	17 170	18 184	17 967	20 625	20 176	12 160	14 207	15 148	16 502	18 748	13 981
Cable covering	46 059	48 008	41 667	39 013	39 395	20 048	14 452	13 705	13 851	16 393	13 408
Caulking lead	31 396	27 209	20 396	18 195	17 907	12 969	11 317	8 725	9 909	8 017	5 684
Casting metals	6 802	6 605	6 476	6 550	6 810	6 995	6 085	5 428	3 611	22 745	19 021
Collapsible tubes	9 900	9 109	3 647	2 594	2 257	2 010	2 113	1 863			
Foil	5 009	4 007	4 166	4 522	3 995	2 908	4 650	3 237	10 479 }	7 186 }	8 597 }
Pipes, siphons, bends	16 228	16 487	16 130	19 315	14 928	12 912	12 509	10 555	12 626	20 432	19 796
Sheet lead	19 096	25 045	21 470	21 223	19 318	22 552	22 170	15 205			
Solder	63 237	63 515	64 672	65 109	60 128	52 022	57 447	58 319	68 390	54 278	41 366
Storage batteries:											
Antimonial lead	257 142	292 328	314 997	331 628	355 144	296 390	348 217	416 702	412 564	375 554	302 240
Lead oxides	281 229	324 380	344 156	366 403	417 670	338 108	397 855	441 383	466 710	438 778	343 117
Type metal	22 204	18 880	18 093	19 887	18 612	14 706	13 614	12 886	10 795	10 019	8 997
White lead	5 385	4 292	2 553	1 587	1 811	2 266	2 715	5 998		26 717	20 736
Red lead and litharge	70 048	56 098	63 321	81 263	87 238	59 382	77 476	70 825	91 642 }	48 758	45 361
Other pigments[‡]	14 138	13 326	15 060	15 821	16 378	10 085	15 599	13 879		15 315	12 333
Tetraethyl lead	252 656	239 715	252 507	248 941	227 252	189 243	217 505	211 292	178 331	186 945	127 903
Other chemicals	566	364	770	856	642	164	132	117			
Miscellaneous uses	35 218	36 884	48 077	46 508	49 055	42 622	55 199	57 290	72 048	45 584	31 293
Total	1 234 272	1 298 648	1 347 401	1 398 161	1 450 976	1 176 707	1 351 771	1 435 472	1 432 744	1 358 335	1 070 303

*Preliminary figures.

[†] According to "Bureau of Mines". Including remelted lead, lead in alloys, lead in scrap and residues and in ores used directly in the manufacture of lead compounds.

[‡] Includes lead content of leaded zinc oxide production.

Table XV. U.S. Secondary Lead Production

Year	Total production (metric tons)	Year	Total production (metric tons)
1965	522 377	1974	633 851
1966	519 669	1975	597 344
1967	502 376	1976	659 136
1968	499 751	1977	757 596
1969	547 856	1978	769 236
1970	541 946	1979	801 368
1971	541 408	1980	673 458
1972	559 370	1981	639 314
1973	593 561		

Although crudeness of the entire exercise must be granted, it is still interesting to extend Calloway's world pool calculations to include data from 1960 to 1980. With the same (or as similar as possible) assumptions, the current world pool estimate is as follows:

$$1980 \text{ World pool} = \frac{\overset{\text{(World consumption)}}{(5\,311\,000)} \times \overset{\text{(U.S. pool)}}{(7\,400\,000)}}{\underset{\text{(U.S. consumption)}}{(1\,070\,300)}} = 36\,720\,000 \text{ short tons}$$

Even if a 20% error is allowed, the world secondary pool of recyclable lead would be an impressive 30–40 million short tons.

It is evident from the foregoing that secondary lead is growing in importance worldwide, and in view of lead's increasing use for automotive and industrial batteries, this trend is likely to continue into the next decade (1995). With both primary and secondary lead resources abundant and economically attractive, sufficient metal to serve world industrial needs for the future seems assured.

Reference

[1]H. M. Calloway, "LEAD: A Materials Survey," U.S. Bureau of Mines Information Circular, U.S. Bureau of Mines, Washington, D.C., 1962.

12
Safe Use of Lead Compounds

In the past several years, lead exposure has been a prime focus of the media and of governmental regulatory agencies. The result has been that lead has become one of the most studied, regulated, and, when properly handled, safe materials known to man.

Lead is toxic if absorbed in excess. However, modern methods of manufacturing and industrial hygiene practices, combined with insistance on good personal hygiene and a program of medical surveillance, should be sufficient to assure that no employee will suffer adverse health effects from industrial overexposure to lead.

Lead Toxicity

Lead is a natural part of our environment. Besides occurring in large quantities as mineral deposits, it exists in all plant and animal tissue, and consequently, it exists in small quantities in our food and drink. Everyone, therefore, is exposed to lead in his daily life and carries in his bones and bloodstream a low level of lead, which is harmless. There has even been some recent research which suggests that lead may be an essential factor in the metabolism of iron. Lead, as it exists naturally in our environment, presents no health hazard to the general population. We are concerned here not with natural exposure, but with any unusual exposure which may occur in industry.

Not all of the lead that enters the body is retained. It is estimated that about 45% of inhaled lead and less than 10% of ingested lead remains in the body. Most of the retained lead is stored in the dense bone, where for all practical purposes, it is inert. A small percentage of the retained lead enters the blood stream, bone marrow, or soft tissue such as the kidney.

Analysis of blood for lead content is generally considered the best measure of an individual's total lead body burden. The normal range of lead in blood among an adult, nonoccupationally exposed population is from 10 to 40 μg of lead per 100 g of whole blood (μg/100 g).

Under normal conditions, the body enters into equilibrium with lead; i.e., approximately the same amount of lead is excreted as is absorbed. However, under conditions of prolonged high exposure, the body is not able to excrete all the absorbed lead and an increase in the amount of lead retained in the blood occurs.

Elevated blood lead levels are not indicative of adverse health effect unless they cross the threshold of toxicity. Occupational health experience indicates that symptoms of lead poisoning do not occur at blood lead levels less than

159

80 $\mu g/100$ g. Although there are individuals who can tolerate much higher blood lead levels, 80 $\mu g/100$ g represents the upper limit of safety for most.

In order to prevent lead poisoning resulting from industrial overexposure to inorganic lead compounds, blood lead levels should be kept below 80 $\mu g/100$ g. In order to achieve this goal, a comprehensive program of industrial hygiene, personal hygiene, and medical surveillance is necessary.

Controlling Lead in the Workplace

Some Basic Principles

Inorganic lead can only be absorbed in two ways: through inhalation and ingestion. It is not absorbed through the skin. Not all of the lead ingested and/or inhaled is absorbed into the bloodstream, and in the case of certain ceramic lead frits which are much less soluble than others, the rate of absorption of biologically available lead particulate is negligible. An International Lead Zinc Research Organization (ILZRO) sponsored animal study confirmed that galena, Pb 349 frit, and lead aluminum bisilicate were not absorbed in the same amount as soluble frits and white lead over the same amount of time. This suggests that workers exposed to the former group of compounds are at minimal risk of undue absorption compared with those exposed to the latter group.

Much of the danger of overexposure to lead in the ceramic industries arises as a result of the dust created by the transporting, mixing, and handling of soluble inorganic lead compounds. This can be avoided by maximum usage of low-solubility frits.

There are several procedures which may be followed in order to decrease the amount of dust created by product handling. Each organization ought to review the options and choose that method or combination of methods which is best suited to its needs. Among the options available are engineering controls, packaging, and product treatment to reduce dusting. Of course, the use of *low-solubility frits* will minimize the need for expensive engineering controls in the ceramic industries.

Engineering Controls

Engineering controls, where feasible, are the most effective means of controlling the amount of dust in the air. Engineering controls are normally of three types: ventilation, isolation, and enclosure. Ventilation may be either general or specific; that is, it may be directed toward a large general area (such as a lunch room) or it may take the form of a suction hood or vent over a specific operation (such as a spray booth). Isolation is exactly as it sounds — the isolating of a lead operation so that the general plant area is not affected by emissions. Enclosure requires the closing off of an operation without isolating it from the rest of the manufacturing process.

The first step in devising an adequate engineering program is the determination of the current levels of lead in the air and the sources of emissions. This can be accomplished by air-monitoring devices which are either stationary sampling mechanisms to determine lead levels in a given location or portable mechanisms attached to an individual worker to determine his or her personal exposure during the course of a day.

Local regulations may establish an air-lead level which must be reached by engineering controls; however, in the absence of such regulations, feasible engineering controls should be implemented to assure the lowest possible level of exposure to lead.

Packaging to Reduce Dust

Employee exposure to lead dust can be reduced through the use of special packaging methods that eliminate the need for employees to hand-mix compounds. One such method is semibulk, reusable containers. Many manufacturers of inorganic lead compounds can ship their materials in heavy-duty reinforced vinyl packages that are permanently attached to pallets. They are collapsible and can be adapted to either gravity or pneumatic unloading. When emptied, the bags are returned to the supplier to be refilled. Other types of semibulk containers, one type known as "sling bins", are frequently used.

A second system which reduces employee exposure to lead dust is batch-weight packaging. With batch-weight packaging, the lead compound is pre-weighed and packaged in the amounts normally used in the ceramics operation. This eliminates the need for the employee to be constantly scooping and weighing up leaded materials.

One of the newer methods of reducing occupational exposure to lead dust is the use of "dustless leads." These are lead compounds that have been pretreated through various processes which render them nondusting. However, since there may be some disadvantages to "dustless leads," they should only be used after a careful analysis of their advantages relative to other methods of control.

Personal protective gear provides a first line in the defense against excessive lead absorption. Respirators have proven to be effective deterrents to the inhalation of air-borne lead. A respirator program should be established, and it should be consistently and uniformly enforced. Firms in the lead industry which have done so (primary and secondary smelters handling many thousands of tons of lead annually) have experienced dramatic drops in the blood lead levels of their employees. A very successful recent addition to the protective gear is the ventilated helmet. This device combines the functions of the industrial hard hat with a transparent face shield with that of the respirator. In this case though, air is pumped through a filter pack, through a channel in the top of the helmet, and passes between the wearer's face and the protective shield. Thus, the wearer has a constant supply of clean, filtered air.

Housekeeping procedures are also extremely important in controlling the amount of lead dust biologically available in the air — *never dry sweep!* Vacuum sweeping is the preferred method, but when this is not feasible, lead-exposed areas should be wet down before being cleaned up. Employees should be required to keep their work station clean, and general cleanups of the entire plant should be conducted periodically. Studies conducted in battery plants with higher potential lead exposure levels than those generally found in the ceramic industries have demonstrated that careful attention to housekeeping procedures can have a greatly beneficial impact upon employee health.

Improved personal hygiene procedures have proven to be extremely effective in protecting workers from undue absorption of lead, particularly from ingestion. Lead dust can accumulate on the hands, face, or clothing of the careless worker where it becomes available for absorption via food, drink, or tobacco. Employees must be instructed to wash thoroughly before eating, drinking, or smoking, and of course, no food, beverage, tobacco products, or cosmetics should be stored, consumed, or used in a lead-exposed area. The employees should have a separate, lead-free area in which to take their breaks and lunch, and ideally, they should not bring their lead dust with them into the lead-free areas. Removable coveralls would make this possible. Finally, employees should shower before leaving work, and lead-contaminated work clothing should be left at the work place.

Employee information and training programs provide the employee with the information he or she needs to be motivated to comply with a company's health protection program. There is much the employee can do to protect himself from the potentially harmful effects of lead. Motivational and educational programs in use throughout the lead industry have demonstrated the value of enlisting this valuable resource, the employee, in the fight to protect his own health.

Medical Surveillance of Ceramics Workers

A comprehensive medical surveillance program for lead-exposed workers is essential, since the early effects of lead overexposure are subtle but reversible. Thus, without a carefully constructed medical surveillance program, an individual might not notice the slight warning signs of incipient lead intoxication and progress past the stage of easy reversibility of symptoms. A medical surveillance program enables one to identify and resolve potential health concerns before they become serious problems.

The first step in a lead-related medical surveillance program is the initial assessment or preemployment physical. The importance of this initial physical assessment cannot be overemphasized. At this time, a sincere effort must be made to identify preexisting conditions that might make it unwise, if not unhealthy, for a potential employee to be exposed to lead. Not only must the obvious information be sought out (present blood lead level, any previous occupational exposure to lead, kidney impairment, etc.), but so also must more subtle influences on lead absorption be identified, such as smoking, drinking, and personal hygiene habits, hobbies, or part-time work that carries lead exposure.

The next step in the medical surveillance of lead-exposed workers is periodic blood lead level testing. Blood lead samples may be drawn by a plant nurse, company physician, or licensed clinic. The samples should then be sent to an approved laboratory for analysis. The laboratory will report the level of lead in blood for the individual tested. This is usually recorded in micrograms (μg) of lead per 100 mL of whole blood. Thus, a report on an individual might read John Doe — 25 μg/100 mL. Blood lead levels of 40–80 μg/100 mL for an occupationally exposed group are not generally considered harmful, although for fertile, and particularly for pregnant women, blood lead levels should not exceed 30 μg/100 mL.

13
International Standards

Ceramics and glassware have been of great utility and esthetic value to man for thousands of years. Over time, techniques and scientific skills have been developed to enhance not only the beauty and utility of the products but also to assure stable glazes with very low heavy-metal release in food acids. There is no doubt that some civilizations, through improper use of heavy-metal chemicals, affected the general population of their times by heavy-metal release from the foodware. Even today, in certain parts of the world where cottage industries exist, the same problem is inherent in the manufacture of ceramic foodware. Because this occurs in remote areas, attention is not drawn to them by society in general. However, in 1960, a case of lead poisoning occurred in the United States that was caused by improperly fired glazed dinnerware.

In order to prevent the repetition of such an occurrence, the International Lead Zinc Research Organization, the Lead Industries Association, and the U.S. Potters Association began a long series of research projects and investigations into the control of lead and cadmium release from ceramic foodware. In these years, the determination of low levels of lead and cadmium was a very difficult procedure. The standard analytical technique was that of utilization of a dithizone test with a titrated end point. Through a cooperative effort among the previously mentioned groups, an ASTM test was developed and was designated as ASTM C-555 (1964). At the same time the United States Food and Drug Administration was developing a similar test for their own purposes of monitoring imported and domestically produced ceramic foodware. Refinements of the techniques continued, and the use of the atomic absorption spectrophotometer replaced the dithizone procedure; this gave greater reproducibility and certainly greater ease of testing. At the same time, it was recognized that the samples to be evaluated were inconsistent with those used by the Food and Drug Administration (FDA). An updating of C-555 was made in order to correspond to the sampling techniques used by the Food and Drug Administration. This led to the adoption by ASTM Committee C21 of a new standard method of test for lead and cadmium extracted from glazed ceramic surfaces which received the designation C738-72. This test consisted of the sample being exposed to 24-h leaching by a 4% acetic acid solution at room temperature. The determination for lead and cadmium release was then made by using atomic absorption spectrophotometry. Limits of lead and cadmium release were adopted as 7 ppm of lead and 0.5 ppm of cadmium. These limit values were consistent with those being used by the FDA at that time. During the decade of the 1960's, the main driving force for developing the cooperative research between the International Lead Zinc Research Organization, Lead Industries Association, the U.S. Potters

Association, and Rutgers University was John S. Nordyke. While working with John Koenig of Rutgers, he developed a program which eventually evolved into the publishing of a book entitled "Lead Glazes for Dinnerware." The wide distribution of this book was responsible for the consolidation of test methods and practices in pottery manufacture which utilized lead and cadmium glazes. It is important to note the basic conclusions that have stemmed from this sponsored research. Nordyke, in his keynote address to the First International Congress on Ceramic Foodware Safety, summarized the results as follows:

"(1) The presence of lead, or other heavy metal compounds in a glaze, does not of itself constitute a hazardous situation. The factor of importance is how resistant to attack by food acids the glaze may be.

(2) Among the most acid resistant of all ceramic glazes are lead-fluxed glazes, properly formulated, properly applied, and properly fired.

(3) Not only are properly made lead-fluxed glazes among the most resistant to attack by food acids, but they are also the most resistant to destructive action by dishwashing detergents."

As more research became available in the literature and as a number of national standards around the world evolved, it became apparent that there was a great deal of inconsistency both in the test method and in the limits of release for heavy metals from ceramic foodware. In May of 1972, the International Lead Zinc Research Organization held a meeting in Montreal, Canada. It served to promote "new ideas" in lead and zinc research. On the basis of the extensive research involvement that Rutgers had with the aforementioned sponsors, M. G. McLaren suggested that an international conference on ceramic foodware safety be held. Its devout purpose would be to attempt to internationalize standards and limits of release for lead and cadmium from foodware. It was also suggested at that meeting that the proper forum for such a conference would be the World Health Organization (WHO) in Geneva, Switzerland. Dodd S. Carr had planned a trip to Geneva immediately following the Montreal conclave. It was suggested that he determine if the World Health Organization would be willing to participate in such a program. His visit to WHO precipitated a meeting with Frank C. Lu, the director of the Food Additives Program for WHO. It was determined that there had been a great deal of interest in contamination of food substances by containing vessels, and consequently, WHO indicated that, while they could not host such an international conference, they would be willing to supply the facilities of WHO for such a purpose.

As a result of these determinations, Rutgers University requested funding from International Lead Zinc Research Organization, Lead Industries Association, and the Food and Drug Administration to convene a meeting of the International Conference on Ceramic Foodware Safety held at World Health Organization's Headquarters in Geneva in Nov. 1974. Over 150 scientists, engineers, compliance officers, and toxicologists were invited to attend this first conference. This represented a worldwide spectrum of the people involved in the manufacture, surveillance, and health aspects of ceramic foodware. Preparation for the meeting involved a number of organizations beyond the sponsors of the program and the

World Health Organization. Ambassador Dale of the U.S. Mission in Geneva was of enormous constructive help in the organization of the meeting, which lasted for 3 days and involved strenuous discussion between various countries, trade organizations, and compliance groups. Each representative felt obliged to put forth his own nation's procedures as the world standard, and it became obvious that there would be great difficulty in reaching agreement on a common world standard that all countries would use. Much could be said about the scientific detail and toxicological reports that were given at this meeting, but the most significant results were the resolutions that were adopted by this international conference on ceramic foodware safety. These may be summarized as follows with the understanding that the term "ceramic" shall include ceramics, glass, vitreous enamels, and glass-ceramics. The following unanimous resolutions were made:

(1) The conference recognized the need to establish limits of lead and cadmium release and internationally accepted uniform methods of analysis to deal with ceramic foodware safety.

(2) The conference recognized the need to establish a correlation between the extraction of lead and cadmium by food from ceramic foodware and a well-defined acetic acid extraction test.

(3) The conference recognized the need for examining the extent to which lead and cadmium extracted from ceramic foodware contributes to the provisional tolerable weekly intake of these elements.

(4) The conference recognized the need to establish an educational program to reach the industry, "cottage industry," art hobby potters, and teachers, with concise information on control of lead and cadmium in ceramic foodware glazes and decorations throughout the world.

(5) The conference recognized the competence of the World Health Organization to determine the proper organizational path to deal with the above matters.

(6) The conference agreed that the participants of the meeting should ask their delegates to the World Health Assembly to request that WHO address itself to the matter of ceramic foodware safety.

Of these recommendations by the meeting participants, the most important was the request that WHO accept the problem of foodware safety as a task. As a result, the U.S. delegate to the World Health Assembly, working in conjunction with Frank C. Lu, and at the request of other countries' assembly delegates, brought the matter before the world assembly. It was recognized that if there were uneven world standards, a barrier to international trade would exist for the ceramic industry. In fact, this was restricting trade between nations at this time. The United Nations and the Gatt General Agreement on Tariff and Trade were both made aware of this restriction, and they supported the World Health Organization in assuming the tasks. In Feb. 1976, Frank C. Lu indicated that the assembly had accepted the task for the World Health Organization and he had been instructed to organize a program to develop a base of understanding for obliterating problems in ceramic foodware safety. At that time, Lu appointed M. G. McLaren to the expert panel on food additives to aid in organizing further meetings on this problem. By utilization of the base of knowledge developed at the 1974 conference and also the spectrum of people from that meeting, a conference was called by WHO in 1976 to review the problem and make recommendations. The importance of this meeting was evident in that it attracted not only the participating countries of the World Health Organization, but also that it had a very sound toxicological base in the pursuing discussions. The recommendations that emanated from the expert panel meeting were in general agreement with the test methods and limits of release already established. This employs a 4% acetic acid test for 24 h of leaching with the maximum levels of release being 7 ppm for flatware, 5 ppm for small hollowware, and 2.5 ppm of lead release for large hollowware. The suggested limits of cadmium release were one-tenth the value of those to be used for lead. The meeting also suggested that there was more work to be done in the development of room-temperature procedures and limits adequately representing the extraction of lead and cadmium from cookware that is intended for use at higher temperatures. It was further recommended that a future WHO meeting on ceramic foodware safety be convened to consider educational methods intended to improve the quality of foodware and thereby reduce health hazards. New test methods for cookware and simple screening tests for better compliance with health and safety standards were also necessary. When the report from the expert panel was received, many countries adopted the procedures as outlined. Coincident with the reports that were generated by WHO, an effort was being undertaken by the International Standards Organization (ISO) in Geneva, Switzerland, to develop ISO standards based on the WHO work for use as international standards. N. N. Chopra had submitted a request to all members of ISO to review their existing national standards and make recommendations for this undertaking by ISO. Technical Committee 166 was formed with the approval of the council of ISO, and the secretariat was awarded to the Standards Institute of Israel, who convened a meeting of this committee in June 1977 to determine the scope and procedures for establishing these standards. At that time, Subcommittee TC166 SC1 was formed. The secretariat of that committee was awarded to the United States, and the responsibility fell to Rutgers University to determine the amount of lead and cadmium release from ceramic foodware in contact with food acids. It was hoped that the test procedures and limits of release as adopted by the World Health Organization could be translated into ISO standards with minimal effort. For this purpose, a meeting of the TC166 SC1 was called in October of 1977 in Philadelphia to attempt to amend the working draft. After several rounds of discussion and voting by the participating countries, the draft was assigned the proposal number of DIS6486 and was submitted to the next convened meeting of the committee of TC166 in Israel in May 1978. During this meeting, the DIS6486 was passed and sent to the membership of ISO for ballot. Also at this meeting, Subcommittee TC166 SC2 for glass and glass-ceramics was created in order to write the international standards pertaining to those materials. Again, the first order of business was the setting of standards for lead and cadmium release from glass and glass-ceramics parallel to the standards which would be developed for the ceramic foodware.

In order to expedite the procedures with ISO, a joint meeting was called by the World Health Organization with the participation of ISO. The meeting was to serve two purposes: (1) that of reviewing the suggestions of the 1976 meeting of WHO and (2) that of giving credence to any limits of release that might be determined by ISO by the assembled toxicologists at the joint meeting. At this WHO meeting, the limits of release and test methods as previously developed in 1976 were reaffirmed with the inclusion of a change for testing flatware. The test would be based on a milligrams per square decimeter release basis instead of a volumetric basis of 7 ppm. This was consistent with the request that was being developed by ISO and had the unanimous support of the participants. Although the WHO meeting was successful in reaffirming test procedure and limits of release, the ISO document DIS6486 failed in its vote and consequently a meeting of the TC166 SC1 was convened in New York in 1980 in order to revise the DIS and send it out for a new vote. In Jan. 1981 a positive vote was received on the DIS, and a new standard for ceramic foodware for lead and cadmium release in contact with food acids was generated as ISO Standard 6486, Part I, Methods of Test, and Part II, Limits of Release for Ceramic Foodware in Contact with Food. The actions of TC166 SC2 developed a similar proposal for glass and glass-ceramic ware at meetings in 1980 and 1982 held in Berlin under the auspices of the Deutsches Institut für Normung, who acted as the secretariat for this technical subcommittee. In 1982 the ballot for DIS6486 passed, and standards were evolved.

As a result of this 20-year activity in research and development of stability of lead and cadmium ions in glasses and glazes, manufacturers have developed better procedures and a greater understanding of the problem. It is also very important to note that there are now existing standards developed by both the World Health Organization and the International Standards Organization for ceramics, glass, and glass-ceramics which have helped to eliminate barriers of trade and provide a greater degree of safety for the consuming public.

The generation of these standards has been almost a unique example of the cooperation of engineers, scientists, compliance personnel, toxicologists, and international organizations. Work is continuing, but in effect, 90% of the standards and test methods are completed. The development of an appropriate test for cookware is proceeding at the time of the writing of this article.

Acknowledgments

The contributors to these programs have been legend, but recognition should be given to a special few. John S. Nordyke recognized the problems early and was instrumental in developing cooperative research between the manufacturers, trade organizations, and the universities. He served as the keynote lecturer at the first conference and has served as chairman of International Standards Organization technical subcommittee SC1 to develop the standards for ceramic foodware. Edward A. Steele of the Food and Drug Administration played a major role in developing the standards in the best interests of the consuming public. Frank C. Lu, who was the director of the Food Additives Program in the World Health Organization, should be recognized for his insight into the development of the WHO recommendations. Dodd S. Carr of International Lead Zinc Research Organization was diligent in his management of the various meetings. M. Parkany of ISO was most helpful in expediting organizational matters. Certain organizations should be cited, such as International Lead Zinc Research Organization, Lead Industries Association, Food and Drug Administration, Rutgers University, World Health Organization, International Standards Organization, the DIN, the Standards Institute of Israel, and Hammond Lead Products, Inc.

ISO 6486 Parts 1 and 2 and ISO 7086 Parts 1 and 2 are reproduced with the permission of the International Organization for Standardization (ISO). Copies of these documents can be obtained from the ISO Central Secretariat, Case postale 56, 1211 GENEVA 20, Switzerland and from the ISO Member Bodies, in the case of the USA from the American National Standards Institute (ANSI), 1430 Broadway, New York, N.Y. 10018 USA.

Bibliography

"Ceramic Foodware Safety, Critical Review of Sampling, Analysis, and Limits For Lead And Cadmium Release," World Health Organization, Geneva, Nov. 12–14, 1979.
"Ceramic Foodware Safety, Sampling Analyses and Results for Release," *WHO Food Addit. Ser.*, **8**, (1976).
"International Conference on Ceramic Foodware Safety," Proceedings of the Lead Industries Association, Inc., New York, 1975.
WHO Food Addit. Ser., **4** (1972).
WHO Tech. Rep. Ser., **505** (1972).

International Standard **ISO** 6486/1

INTERNATIONAL ORGANIZATION FOR STANDARDIZATION•МЕЖДУНАРОДНАЯ ОРГАНИЗАЦИЯ ПО СТАНДАРТИЗАЦИИ•ORGANISATION INTERNATIONALE DE ŇORMALISATION

Ceramic ware in contact with food —
Release of lead and cadmium —
Part 1 : Method of test

Articles en céramique en contact avec les aliments — Émission de plomb et de cadmium — Partie 1 : Méthode d'essai

First edition — 1981-06-01

UDC 642.72 : 666.5 : 620.1

Ref. No. ISO 6486/1-1981 (E)

Descriptors : ceramics, earthenware, tableware, chemical analysis, determination of content, toxic substances, lead, cadmium, composition tolerances.

Foreword

ISO (the International Organization for Standardization) is a worldwide federation of national standards institutes (ISO member bodies). The work of developing International Standards is carried out through ISO technical committees. Every member body interested in a subject for which a technical committee has been set up has the right to be represented on that committee. International organizations, governmental and non-governmental, in liaison with ISO, also take part in the work.

Draft International Standards adopted by the technical committees are circulated to the member bodies for approval before their acceptance as International Standards by the ISO Council.

International Standard ISO 6486/1 was developed by Technical Committee ISO/TC 166, *Ceramic ware, glassware and glass ceramic ware in contact with food*, and was circulated to the member bodies in June 1979. It results from the division into two parts of ISO/DIS 6486.

It has been approved by the member bodies of the following countries :

Austria	Italy	South Africa, Rep. of
Brazil	Japan	Thailand
Canada	Korea, Rep. of	United Kingdom
Czechoslovakia	Philippines	USA
Germany, F. R.	Poland	
Israel	Romania	

The member bodies of the following countries expressed disapproval of the document on technical grounds :

Denmark
Ireland

Ceramic ware in contact with food —
Release of lead and cadmium —
Part 1 : Method of test

0 Introduction

The problem of lead and cadmium release from ceramic ware requires effective means of control to ensure the protection of the population against possible hazards arising from the use of improperly formulated, applied and fired glazes and decorations on the food contact surfaces of ceramic ware used for the preparation, serving and storage of food and beverages. As a secondary consideration, different requirements from country to country for the control of the release of toxic materials from the surfaces of ceramic ware present non-tariff barriers to international trade in these commodities. Accordingly, there is a need to establish internationally accepted methods of testing ceramic ware for lead and cadmium release, and to define permissible limits for the release of these toxic heavy metals.

An expert panel convened by the World Health Organization (WHO) met in Geneva, in June 1976, and recommended the adoption of sampling methods, testing procedures and limits for the release of toxic materials from ceramic ware. A further meeting was convened by WHO in November 1979. The method of test specified in this International Standard is based on the WHO recommendations[1, 2, 3, 4, 5].

1 Scope

This part of ISO 6486 specifies a method of test for the release of lead and cadmium by ceramic ware which may be used in contact with food, for example ceramic ware made of china, porcelain and earthenware, whether glazed or not, but excluding glass, glass ceramic and porcelain enamel articles.

2 Field of application

This part of ISO 6486 is applicable to ceramic ware which may be used for the preparation, serving and storage of food and beverages, excluding articles used in food manufacturing industries or those in which food is sold.

3 Reference

ISO 3585, *Glass plant, pipeline and fittings — Properties of borosilicate glass 3.3.*

4 Definitions

For the purpose of this part of ISO 6486, the following definitions apply.

4.1 ceramic ware : Ceramic articles which may be used in contact with foodstuffs, for example foodware made of china, porcelain and earthenware, whether glazed or not.

4.2 flatware : Ceramic ware having an internal depth not exceeding 25 mm, measured from the lowest point to the horizontal plane passing through the point of overflow.

4.3 hollow-ware : Ceramic ware having an internal depth greater than 25 mm, measured from the lowest point to the horizontal plane passing through the point of overflow.

Hollow-ware may be termed large or small according to its capacity as follows :

    a) large hollow-ware : hollow-ware with a capacity of 1,1 litres or more;

    b) small hollow-ware : hollow-ware with a capacity of less than 1,1 litres.

4.4 test solution : The solvent used in the test to extract lead and cadmium from the ceramic ware.

5 Principle

Extraction of lead and cadmium by an acetic acid solution from the ceramic ware surfaces that would normally come into contact with food. Determination by atomic absorption spectrometry of the amounts of lead and cadmium extracted.

6 Reagents

All reagents shall be of recognized analytical grade. Distilled water or water of equivalent purity shall be used throughout.

6.1 Acetic acid (CH$_3$COOH), glacial, ϱ = 1,05 g/ml.

Store this reagent in darkness.

6.2 Test solution : acetic acid, 4 % (V/V) solution.

Add 40 ml of the glacial acetic acid (6.1) to distilled water, and dilute to 1 000 ml. This solution shall be freshly prepared for use.

6.3 Analytical stock solutions

Prepare analytical stock solutions containing 1 000 mg of lead per litre and at least 500 mg of cadmium per litre in the acetic acid solution (6.2) or in a 2 % (V/V) nitric acid solution.

Alternatively, appropriate, commercially available, standardized lead and cadmium AAS solutions may be used.

7 Apparatus

7.1 Atomic absorption spectrometer, having a minimum sensitivity of 0,50 mg of lead per litre, and 0,05 mg of cadmium per litre, for 1 % absorption. It shall be operated in accordance with the manufacturer's instructions. A digital concentration reader (DCR) is optional but useful for rapid analysis.

7.2 Glassware, of borosilicate glass, as specified in ISO 3585.

8 Sampling

8.1 Priority

Carry out sampling of ceramic ware in the following order of priority :

   a)   large hollow-ware;

   b)   small hollow-ware;

   c)   flatware.

Articles having the highest surface area/volume ratio within each category should be given preference. Articles that are highly coloured or decorated on their food contact surfaces should be especially considered for sampling.

8.2 Sample size

It is desirable to develop a system of control that is regarded as appropriate to the circumstances. If available, six articles shall be tested. Each of the articles shall be identical in size, shape, colour and decoration.

8.3 Preparation and preservation of test samples

Samples of dinnerware shall be clean and free from grease or other matter likely to affect the test.

Briefly wash the specimens at a temperature of about 40 °C with a solution containing a non-acidic detergent. Rinse in tap water and then in distilled water or water of equivalent purity. Drain, and dry in either a drying oven or by means of a new filter paper so as to avoid any stains. Do not handle the surfaces to be tested after cleaning.

9 Procedure

9.1 Determination of filling volume

Place each specimen on a flat horizontal surface and fill it with water to 5 mm from overflowing, as measured along the surface of the specimen. Measure the volume (V) of the water to an accuracy of ± 2 %.

9.2 Determination of reference surface area for flatware

Invert the specimen on graph paper marked in millimetre squares and draw the contour round the rim. Calculate the area enclosed by the contour and record this as the reference surface area (A_R) in square decimetres.

9.3 Extraction

9.3.1 Extraction temperature

Carry out the extraction at a temperature of 22 ± 2 °C.

9.3.2 Leaching

Fill each specimen with the test solution (6.2) to 5 mm from overflowing, as measured along the surface of the specimen. Cover the specimen to prevent exposure of the surface under test to light. Leach for 24 h ± 10 min.

9.4 Sampling of the extraction solution for analysis

Prior to sampling the extraction solution to determine the lead and/or cadmium concentration, mix the extraction solution of each specimen by an appropriate method which avoids any loss of extraction solution or any abrasion of the surface being tested (for example, using a pipette, remove and allow the extraction solution to run back onto, and into, the specimen several times). Do not dilute the extraction solution (for example by rinsing the specimen). Transfer the extraction solution to a suitable storage container made of borosilicate glass. It is not necessary to transfer all the extraction solution.

Analyse the extraction solution as soon as possible as there is a risk of adsorption of lead or cadmium onto the walls of the storage container, particularly when the metals are present in low concentrations.

9.5 Calibration

Establish and carefully standardize instrument operating techniques so as to utilize maximum sensitivity, as determinations of lead concentrations as low as 0,50 mg/l, or cadmium concentrations as low as 0,05 mg/l, require the full potential of most instruments.

Prepare standard solutions by diluting the analytical stock solutions (6.3) with the test solution (6.2), and use the bracketing technique or construct a calibration curve having, for example, the absorbances of the standard solutions as abscissae, and the corresponding lead or cadmium contents, in milligrams per litre, as ordinates. Carry out a blank test on the reagents used for each set of determinations.

9.6 Determination of lead and cadmium

Determine the approximate amount of lead and cadmium in the extraction solution by use of the bracketing technique using the prepared standard solutions (see 9.5). This procedure may be used with any available read-out device. An averaging device, if available on the read-out device, will reduce the effects of background "noise" and improve both accuracy and precision.

If the lead content of the extraction solution is found to be higher than 20 mg/l, take a suitable aliquot portion and dilute it with the test solution (6.2) to reduce the concentration to less than 20 mg/l.

Alternatively, standard solutions of higher concentration may be used.

Similar considerations apply to the determination of cadmium.

Determine the lead and cadmium contents of the extraction solution by atomic absorption spectrometry using the procedure specified by the instrument manufacturer.

10 Expression of results

10.1 Bracketing technique

The lead or cadmium content, c_0, expressed in milligrams per litre of the extraction solution, is given by the formula

$$\frac{A_0 - A_1}{A_2 - A_1} \times (c_2 - c_1) + c_1$$

where

A_0 is the absorbance of the lead or cadmium in the extraction solution;

A_1 is the absorbance of the lead or cadmium in the lower bracketing solution;

A_2 is the absorbance of the lead or cadmium in the upper bracketing solution;

c_1 is the lead or cadmium content, in milligrams per litre, of the lower bracketing solution;

c_2 is the lead or cadmium content, in milligrams per litre, of the upper bracketing solution.

NOTE — If the extraction solution was diluted (see 9.6), an appropriate correction factor has to be used in the formula.

10.2 Calibration curve technique

Read the lead or cadmium content directly from the calibration curve or from the direct read-out.

10.3 Calculation of release of lead and cadmium from flatware

The lead or cadmium released per unit area from flatware, a_0, expressed in milligrams per square decimetre, is given by the formula

$$\frac{c_0 \, V}{A_R}$$

where

c_0 is the lead or cadmium content, expressed in milligrams per litre, of the extraction solution (10.1 or 10.2);

V is the filling volume, expressed in litres, of the specimen (9.1);

A_R is the reference surface area, expressed in square decimetres, of the specimen (9.2).

10.4 Reporting

For hollow-ware, report the result to the nearest 0,1 mg of lead per litre and to the nearest 0,01 mg of cadmium per litre.

For flatware, report the result to the nearest 0,1 mg of lead per square decimetre and to the nearest 0,01 mg of cadmium per square decimetre.

11 Test report

The test report shall include the following particulars :

a) a reference to this part of ISO 6486;

b) identification of the sample;

c) the results and the method of expression used;

d) any unusual features noted during the determination;

e) any operation not included in this International Standard, or regarded as optional.

173

Bibliography

[1]   *Proceedings, International Conference on Ceramic Foodware Safety,* pp. 8-17, 1975, Lead Industries Association Inc.,
      292 Madison Avenue, New York, N.Y. 10017, USA.

[2]   WHO Food Additives Series No. 4, 1972

[3]   WHO Technical Report Series No. 505, 1972.

[4]   WHO/Food Additives 77.44. *Ceramic Foodware Safety, Sampling, Analysis and Limits for release* (Report of a WHO Meeting,
      Geneva 8-10 June 1976).

[5]   WHO/Food Additives HCS/79.7. *Ceramic Foodware Safety, Critical Review of Sampling, Analysis, and Limits for Lead and
      Cadmium Release* (Report of a WHO Meeting, Geneva 12-14 November 1979).

International Standard

ISO 6486/2

INTERNATIONAL ORGANIZATION FOR STANDARDIZATION●МЕЖДУНАРОДНАЯ ОРГАНИЗАЦИЯ ПО СТАНДАРТИЗАЦИИ●ORGANISATION INTERNATIONALE DE ÑORMALISATION

Ceramic ware in contact with food — Release of lead and cadmium —
Part 2 : Permissible limits

Articles en céramique en contact avec les aliments — Emission de plomb et de cadmium — Partie 2 : Limites admissibles

First edition — 1981-08-01

UDC 642.72 : 666.5 : 614.3

Ref. No. ISO 6486/2-1981 (E)

Descriptors : ceramics, earthenware, tableware, chemical analysis, determination of content, toxic substances, lead, cadmium, composition tolerances.

Foreword

ISO (the International Organization for Standardization) is a worldwide federation of national standards institutes (ISO member bodies). The work of developing International Standards is carried out through ISO technical committees. Every member body interested in a subject for which a technical committee has been set up has the right to be represented on that committee. International organizations, governmental and non-governmental, in liaison with ISO, also take part in the work.

Draft International Standards adopted by the technical committees are circulated to the member bodies for approval before their acceptance as International Standards by the ISO Council.

International Standard ISO 6486/2 was developed by Technical Committee ISO/TC 166, *Ceramic ware, glassware and glass ceramic ware in contact with food*, and was circulated to the member bodies in June 1979. It results from division into two parts of ISO/DIS 6486.

It has been approved by the member bodies of the following countries :

Austria	Japan	South Africa, Rep. of
Brazil	Korea, Rep. of	Thailand
Canada	Philippines	United Kingdom
Israel	Poland	USA
Italy	Romania	

The member bodies of the following countries expressed disapproval of the document on technical grounds :

Czechoslovakia
Denmark
Germany, F. R.*
Ireland

---

* Germany, F. R. disapproved only the permissible values 1,7 and 0,17 mg/dm^2 for lead and cadmium of flatware, because of the expected Directive of the European Communities in this field.

Ceramic ware in contact with food — Release of lead and cadmium — Part 2 : Permissible limits

0 Introduction

The problem of lead and cadmium release from ceramic ware requires effective means of control to ensure the protection of the population against possible hazards arising from the use of improperly formulated, applied and fired glazes and decorations on the food contact surfaces of ceramic ware used for the preparation, serving and storage of food and beverages. As a secondary consideration, different requirements from country to country for the control of the release of toxic materials from the surfaces of ceramic ware present non-tariff barriers to international trade in these commodities. Accordingly, there is a need to establish internationally accepted methods of testing ceramic ware for lead and cadmium release, and to define permissible limits for the release of these toxic heavy metals.

An expert panel convened by the World Health Organization (WHO) met in Geneva, in June 1976, and recommended the adoption of sampling methods, testing procedures and limits for the release of toxic materials from ceramic ware. A further meeting was convened by WHO in November 1979. The permissible limits specified in this International Standard are based on the WHO recommendations.[1, 2, 3, 4, 5] As the capability of the industry increases, efforts will be made to reduce these limits for lead and cadmium release.

1 Scope

This part of ISO 6486 specifies permissible limits for the release of lead and cadmium by ceramic ware intended for use in contact with food, for example ceramic ware made of china, porcelain, and earthenware, whether glazed or not, but excluding glass, glass ceramic and porcelain enamelled articles.

ISO 6486/1[1] specifies a method of test for the release of lead and cadmium by ceramic ware which may be used in contact with food.

2 Field of application

This part of ISO 6486 is applicable to ceramic ware intended to be used for the preparation, serving and storage of food and beverages, excluding articles used in food manufacturing industries. Ceramic ware for packaging is also excluded, except when the ware is intended to be retained and used by the purchaser as ceramic ware as defined in this part of ISO 6486. Ceramic ware as defined in this part of ISO 6486 and excluded from its scope should comply with extraction limits that are no less strict than those for similar articles used in the home or in eating establishments, but it may be necessary for stricter standards to apply, such as statutory limits for contaminants in food for sale. (It must be stressed that compliance with the specified extraction limits is not an alternative to compliance with such statutory limits for food.)

3 Definitions

For the purpose of this part of ISO 6486, the following definitions apply.

3.1 ceramic ware : Ceramic articles intended for use in contact with foodstuffs, for example foodware made of china, porcelain and earthenware, whether glazed or not.

3.2 flatware : Ceramic ware having an internal depth not exceeding 25 mm, measured from the lowest point to the horizontal plane passing through the point of overflow.

3.3 hollow-ware : Ceramic ware having an internal depth greater than 25 mm, measured from the lowest point to the horizontal plane passing through the point of overflow.

Hollow-ware may be termed large or small according to its capacity as follows :

a) large hollow-ware : hollow-ware with a capacity of 1,1 litres or more;

b) small hollow-ware : hollow-ware with a capacity of less than 1,1 litres.

4 Permissible limits

The permissible limits for lead and cadmium release from any individual article, when determined by the method described in ISO 6486/1, shall not exceed the values given in the following table.

Type of ceramic ware	Unit	Lead	Cadmium
Flatware	mg/dm^2	1,7	0,17
Small hollow-ware	mg/l	5,0	0,50
Large hollow-ware	mg/l	2,5	0,25

---

1) ISO 6486/1, *Ceramic ware in contact with food — Release of lead and cadmium — Part 1 : Method of test.*

Bibliography

[1]   *Proceedings, International Conference on Ceramic Foodware Safety,* pp. 8-17, 1975, Lead Industries Association Inc.,
       292 Madison Avenue, New York, N.Y. 10017, USA.

[2]   WHO Food Additives Series No. 4, 1972.

[3]   WHO Technical Report Series No. 505, 1972.

[4]   WHO/Food Additives 77.44. *Ceramic Foodware Safety, Sampling, Analysis and Limits for release* (Report of a WHO Meeting,
       Geneva, 8-10 June 1976).

[5]   WHO/Food Additives HCS/79.7. *Ceramic Foodware Safety, Critical Review of Sampling, Analysis, and Limits for Lead and
       Cadmium Release* (Report of a WHO Meeting, Geneva, 12-14 November 1979).

International Standard

ISO 7086/1

INTERNATIONAL ORGANIZATION FOR STANDARDIZATION•МЕЖДУНАРОДНАЯ ОРГАНИЗАЦИЯ ПО СТАНДАРТИЗАЦИИ•ORGANISATION INTERNATIONALE DE NORMALISATION

Glassware and glass ceramic ware in contact with food — Release of lead and cadmium — Part 1 : Method of test

Articles en verre et en vitro céramique en contact avec les aliments — Émission de plomb et de cadmium — Partie 1 : Méthode d'essai

First edition — 1982-11-15

UDC 666.172.3/.5 : 615.9

Ref. No. ISO 7086/1-1982 (E)

Descriptors : glassware, tableware, tests, determination of content, lead, cadmium, definitions, sampling.

Foreword

ISO (the International Organization for Standardization) is a worldwide federation of national standards institutes (ISO member bodies). The work of developing International Standards is carried out through ISO technical committees. Every member body interested in a subject for which a technical committee has been set up has the right to be represented on that committee. International organizations, governmental and non-governmental, in liaison with ISO, also take part in the work.

Draft International Standards adopted by the technical committees are circulated to the member bodies for approval before their acceptance as International Standards by the ISO Council.

International Standard ISO 7086/1 was developed by Technical Committee ISO/TC 166, *Ceramic ware, glassware and glass ceramic ware in contact with food*, and was circulated to the member bodies in May 1981.

It has been approved by the member bodies of the following countries :

Austria	France	Poland
Brazil	Germany, F. R.	Romania
Canada	Ireland	South Africa, Rep. of
Czechoslovakia	Israel	Spain
Denmark	Japan	United Kingdom
Egypt, Arab Rep. of	Mexico	USA

No member body expressed disapproval of the document.

Glassware and glass ceramic ware in contact with food — Release of lead and cadmium — Part 1 : Method of test

0 Introduction

The problem of lead and cadmium release from glassware and glass ceramic ware requires effective means of control to ensure the protection of the population against possible hazards arising from the use of improperly formulated, applied or fired glazes and/or decorations on the food contact surfaces of glassware and glass ceramic ware used for the preparation, serving and storage of food and drinks. As a secondary consideration, different requirements from country to country for the control of the release of toxic materials from the surfaces of glassware and glass ceramic ware present non-tariff barriers to international trade in these commodities. Accordingly, there is a need to establish internationally accepted methods of testing glassware and glass ceramic ware for lead and cadmium release.

An expert panel convened by the World Health Organization (WHO) met in Geneva, in June 1976, and recommended the adoption of sampling methods, testing procedures and limits for the release of toxic materials from ceramic ware.[1] A further meeting was convened by WHO in November 1979.[2]

The method of test specified in this International Standard is based on the WHO recommendations, because it was the sense of the WHO meeting that the term "ceramic" includes ceramics, glass, vitreous enamels, and glass ceramics.

It has been proved that the amount of lead and/or cadmium determined by the method of test specified in this International Standard will not be less than, and in the vast majority of cases will be greater than, the quantities released into acidic foods and drinks over a period of time.[3]

The results of an international survey showed that cooking ware made from glass or glass ceramics is not normally decorated on the food contact surfaces. For that reason this International Standard does not address cooking ware.

1 Scope

This part of ISO 7086 specifies a simulative method of test for the release of lead and cadmium from glassware and glass ceramic ware which may be used in contact with food (including drinks).

2 Field of application

This part of ISO 7086 is applicable to glassware and glass ceramic ware which may be used for the preparation, serving and storage of food.

It does not necessarily apply to glassware made from borosilicate glass or soda-lime-silicate glass, which is not glazed or decorated on any food contact surface, nor need it apply to glass ceramic ware which is not glazed or decorated on any food contact surface.

It does not apply to vitreous and porcelain enamel ware, nor to ceramic ware.

3 References

ISO 385/2, *Laboratory glassware — Burettes — Part 2 : Burettes for which no waiting time is specified.*[4]

ISO 648, *Laboratory glassware — One-mark pipettes.*

---

1) See WHO/Food Additives 77.44. *Ceramic Foodware Safety, Sampling, Analysis and Limits for Release* (Report of a WHO meeting, Geneva, 8-10 June 1976).

2) See WHO/Food Additives HCS/79.7. *Ceramic Foodware Safety, Critical Review of Sampling, Analysis and Limits for Lead and Cadmium Release* (Report of a WHO meeting, Geneva, 12-14 November 1979).

3) Frey, E., Scholze, H. *Blei- und Cadmiumlässigkeit von Schmelzfarben, Glasuren und Emails in Kontakt mit Essigsäure und Lebensmitteln und unter Lichteinwirkung* (Lead and cadmium release from fused colours, glazes and enamels in contact with acetic acid and foodstuffs and under the influence of light). *Ber. Dt. Keram. Ges.* 56 (1979) No. 10, pp. 293-297.

4) At present at the stage of draft. (Partial revision of ISO/R 385-1964.)

ISO 835/2, *Laboratory glassware — Graduated pipettes — Part 2 : Pipettes for which no waiting time is specified.*

ISO 1042, *Laboratory glassware — One-mark volumetric flasks.*

ISO 3585, *Glass plant, pipeline and fittings — Properties of borosilicate glass 3.3.*

ISO 4788, *Laboratory glassware — Graduated measuring cylinders.*

ISO 7086/2, *Glassware and glass ceramic ware in contact with food — Release of lead and cadmium — Part 2 : Permissible limits.*

4 Definitions

For the purpose of this International Standard, the following definitions apply.

4.1 glass : An inorganic, non-metallic material produced by the complete fusion of raw materials at high temperature into a homogeneous liquid which is then cooled to a rigid condition, essentially without crystallization.

4.2 glass ceramic : An inorganic, non-metallic material produced by the complete fusion of raw materials at high temperatures into a homogeneous liquid which is then cooled to a rigid condition with a certain degree of crystallization. It may be translucent or opaque.

4.3 borosilicate glass : A glass containing a sufficient amount of boric oxide to influence its properties, in particular producing high chemical and thermal resistances.

Lead and cadmium are present only in trace amounts as adventitious impurities. The release of these elements will be below the limits of detection of the method of test specified in this International Standard.

4.4 soda-lime-silicate glass : A glass in which the main constituents are normally sodium oxide, calcium oxide, and silica.

Lead and cadmium are present only in trace amounts as adventitious impurities. The release of these elements will be below the limits of detection of the method of test specified in this International Standard.

4.5 foodware : Articles produced from glass and glass ceramics which are intended to be used for the preparation, cooking, serving, and storage of food or drinks, including packaging.

4.6 flatware : Articles having an internal depth not exceeding 25 mm, measured from the lowest internal point to the horizontal plane passing through the point of overflow.

4.7 hollow-ware : Articles having an internal depth greater than 25 mm, measured from the lowest internal point to the horizontal plane passing through the point of overflow.

Hollow-ware may be termed large or small according to its capacity (filling volume, see 8.3.1) as follows :

a) large hollow-ware : hollow-ware with a capacity of 1,1 litres or more;

b) small hollow-ware : hollow-ware with a capacity of less than 1,1 litres.

5 Principle

Filling test specimens with 4 % (V/V) acetic acid solution and keeping them for 24 h at 22 °C in the absence of light.

This solution extracts lead and/or cadmium, if present, from the surfaces of the test specimens.

Determination of the amounts of lead and/or cadmium extracted by atomic absorption spectrometry (AAS).

6 Reagents

All reagents shall be of recognized analytical grade. Unless otherwise specified, distilled water or water of equivalent purity shall be used throughout.

6.1 Acetic acid (CH_3COOH), glacial, $\varrho = 1{,}05$ g/ml.

Store this reagent in darkness.

6.2 Test solution : acetic acid, 4 % (V/V) solution.

Add 40 ml of the glacial acetic acid (6.1) to water, and dilute to 1 000 ml. This solution shall be freshly prepared for use.

6.3 Lead carbonate ($PbCO_3$) **or lead acetate trihydrate** [$Pb(CH_3COO)_2.3H_2O$].

NOTE — Commercially available standard solutions may also be used. (See the note to 6.4).

6.4 Lead, standard solution corresponding to 1 g of Pb per litre.

Dissolve 1,289 6 g of the lead carbonate (6.3), or 1,830 8 g of the lead acetate (6.3), in 40 ml of the glacial acetic acid (6.1) in a 400 ml beaker (7.6). Warm gently to dissolve, then cool the solution and transfer it quantitatively to a 1 000 ml one-mark volumetric flask (7.3). Dilute to the mark with water and mix.

Determine the exact concentration of the solution by a recognized standardized procedure, such as a complexometric titration.

1 ml of this standard solution contains 1 mg of lead.

NOTE — Alternatively, an appropriate, commercially available, standardized lead solution for atomic absorption spectrometry may be used. Prepare the standard solution (6.4) by diluting, as appropriate, with the test solution (6.2) or with 2 % (V/V) nitric acid (HNO_3) solution.

6.5   Lead, standard solution corresponding to 0,1 g of Pb per litre.

By means of a pipette (7.4), transfer 10 ml of the standard lead solution (6.4) to a 100 ml one-mark volumetric flask (7.3), make up to the mark with the test solution (6.2) and mix well. Renew this solution every four weeks.

1 ml of this standard solution contains 0,1 mg of lead.

6.6   Lead, standard matching solutions for calibration.

By means of a burette (7.7), or a graduated pipette (7.5), transfer 0 — 0,5 — 1,0 — 2,0 — 5,0 and 10,0 ml aliquot portions of the standard lead solution (6.5) into separate 100 ml one-mark volumetric flasks (7.3), dilute each to the mark with the test solution (6.2) and mix. These solutions have lead concentrations of 0 — 0,5 — 1,0 — 2,0 — 5,0 and 10,0 mg/l respectively. These solutions shall be freshly prepared for use.

6.7   Cadmium oxide (CdO).

NOTE — Commercially available standard solutions may also be used. (See the note to 6.8).

6.8   Cadmium, standard solution corresponding to 1 g of Cd per litre.

Dissolve 1,142 3 g of the cadmium oxide (6.7) in 40 ml of the glacial acetic acid (6.1) in a 400 ml beaker (7.6). Warm gently to dissolve, then cool the solution and transfer it quantitatively to a 1 000 ml one-mark volumetric flask (7.3). Dilute to the mark with water and mix.

Determine the exact concentration of the solution by a recognized standardized procedure such as a complexometric titration.

1 ml of this standard solution contains 1 mg of cadmium.

NOTE — Alternatively, an appropriate, commercially available standardized cadmium solution for atomic absorption spectrometry may be used. Prepare the standard solution (6.8) by diluting, as appropriate, with the test solution (6.2) or with 2 % (V/V) nitric acid (HNO_3) solution.

6.9   Cadmium, standard solution corresponding to 0,01 g of Cd per litre.

By means of a pipette (7.4), transfer 10 ml of the standard cadmium solution (6.8) into a 1 000 ml one-mark volumetric flask (7.3), make up to the mark with the test solution (6.2) and mix well. Renew this solution every four weeks.

1 ml of this standard solution contains 0,01 mg of cadmium.

6.10   Cadmium, standard matching solutions for calibration.

By means of a burette (7.7), or a graduated pipette (7.5), transfer 0 — 1,0 — 2,0 — 5,0 — 10,0 and 20,0 ml aliquot portions of the standard cadmium solution (6.9) into separate 100 ml one-mark volumetric flasks (7.3), dilute each to the mark with the test solution (6.2) and mix. These solutions have cadmium concentrations of 0 — 0,1 — 0,2 — 0,5 — 1,0 and 2,0 mg/l respectively. These solutions shall be freshly prepared for use.

7   Apparatus

Laboratory glassware shall comply with the requirements of the appropriate International Standards, wherever such International Standards are available. It shall be made of borosilicate glass, as specified in ISO 3585.

Usual laboratory apparatus, and in particular

7.1   Atomic absorption spectrometer, having a minimum sensitivity of 0,50 mg of lead per litre, and 0,05 mg of cadmium per litre, for 1 % absorption. It shall be operated in accordance with the manufacturer's instructions. A digital concentration reader (DCR) attachment is optional, but is useful for rapid analysis.

7.2   Line sources for lead and cadmium.

7.3   One-mark volumetric flasks, of capacities 100 and 1 000 ml, complying with the requirements of ISO 1042, class A.

7.4   One-mark pipettes, of capacities 10 and 100 ml, complying with the requirements of ISO 648, class A.

7.5   Graduated pipettes, of capacities 10 and 25 ml, complying with the requirements of ISO 835/2, class A.

7.6   Beakers.

7.7   Burette, of capacity 25 ml, graduated in divisions of 0,05 ml, complying with the requirements of ISO 385/2, class A.

7.8   Watch-glasses, of different sizes, for covering the test specimens during the test.

7.9   Graduated measuring cylinder, of capacity 500 ml complying with the requirements of ISO 4788.

7.10   Opaque devices, of a suitable shape for covering opaque specimens during the test.

8   Sampling and preparation of test specimens

8.1   Priority

Articles which are highly coloured or decorated on their food contact surfaces or which have a high surface area/volume ratio should be especially selected for testing.

183

8.2 Sample size

It is desirable to develop a system of control that is regarded as appropriate to the circumstances. If available, six articles shall be tested. Each of the articles (test specimens) shall be identical in size, shape, colour and decoration.

8.3 Preparation of test specimens

8.3.1 Determination of filling volume

Select one from the group of identical test specimens, place it on a flat, horizontal surface and fill it with water to 5 mm from overflowing, as measured along the surface of the specimen. Measure and record the volume V of water to an accuracy of ± 2 %.

8.3.2 Determination of reference surface area for flatware

Invert the specimen on graph paper marked in millimetre squares and draw the contour round the rim. Calculate the area enclosed by the contour and record this as the reference surface area S_R in square decimetres to two decimal places. For circular articles the reference surface area may be calculated from the diameter of the specimens.

8.3.3 Preparation of articles which cannot be filled

Articles which cannot be filled to 5 mm from overflowing as specified in 8.3.1 shall be regarded as non-fillable. These articles shall be coated on all surfaces except the reference surface with beeswax or paraffin wax and tested as specified in 9.1.2.2.

8.4 Cleaning the specimens

The specimens shall be clean and free from grease or other matter likely to affect the test results.

Briefly wash them at a temperature of about 40 °C with a solution containing a non-acidic detergent. Rinse in tap water and then in distilled water. Drain and dry either in a drying oven or by wiping with a new filter paper to avoid any stains. Do not handle the surface to be tested after it has been cleaned.

Articles which cannot be filled and which are protected according to 8.3.3 with wax shall be cleaned on the non-protected surface by the same procedure, but shall not dried in an oven.

9 Procedure

9.1 Extraction

9.1.1 Test temperature

Carry out the extraction at a temperature of 22 ± 2 °C; the test solution and the specimens to be tested shall be allowed to attain this temperature before extraction is commenced.

9.1.2 Filling the specimens

9.1.2.1 Place the specimens on a flat, horizontal surface. Add a volume of the test solution (6.2) equal to the filling volume (see 8.3.1), using the measuring cylinder (7.9).

If the specimens are opaque, cover them with a suitable opaque inert material to avoid contamination. It is not necessary to carry out the extraction of such specimens in the dark.

If the specimens are transparent or translucent, cover them at once and place them in the dark.

9.1.2.2 Place specimens of articles which are non-fillable in a borosilicate glass vessel of suitable size and add the test solution (6.2) to completely cover the specimen. Measure and record the required volume V of test solution to an accuracy of ± 2 %. Cover the vessel with a watch-glass (7.8) and place it in the dark.

9.1.3 Duration of extraction

Allow the specimens to stand for 24 h $\pm$ 10 min.

9.2 Sampling the extraction solution for analysis

Prior to sampling the extraction solution to determine the lead and/or cadmium concentration(s), mix the extraction solution from each specimen by an appropriate method which avoids any loss of solution or any abrasion of the surface being tested (for example, using a pipette, remove and allow the extraction solution to run back on to, and into, the specimen several times). Do not dilute the extraction solution (for example by rinsing the specimen).

Transfer the extraction solution to a suitable storage container. It is not necessary to transfer all the extraction solution.

Analyse the extraction solution as soon as possible as there is a risk of adsorption of lead or cadmium on to the walls of the storage container, particularly when the metals are present in low concentrations.

9.3 Calibration

Establish and carefully standardize instrument operating techniques so as to utilize maximum sensitivity, as determinations of lead concentrations as low as 0,50 mg/l, or cadmium concentrations as low as 0,05 mg/l, require the full potential of most instruments (low noise levels).

Determine the absorbances of the standard matching lead solutions (6.6) or the standard matching cadmium solutions (6.10) and, for the determination, either use the bracketing technique or construct calibration curves having, for example, the absorbances of the standard matching solutions as abscissae and the corresponding lead or cadmium concentrations, in milligrams per litre, as ordinates.

Carry out a blank test on the reagents used for each set of determinations.

9.4 Determination of lead and/or cadmium

Determine the lead and/or cadmium concentrations of the extraction solutions by atomic absorption spectrometry following the instrument manufacturer's instructions.

If the lead concentration of the extraction solution is found to be higher than 20 mg/l, or the cadmium concentration higher than 2,0 mg/l, take a suitable aliquot portion and dilute it with the test solution (6.2) to reduce the concentration to less than 20 mg/l for lead or 2,0 mg/l for cadmium.

Alternatively, use standard matching solutions of higher concentrations for the bracketing measurements or for preparing new calibration curves.

10 Expression of results

10.1 Bracketing technique

The lead or cadmium concentration, c_0, expressed in milligrams per litre of extraction solution, is given by the equation

$$c_0 = \frac{A_0 - A_1}{A_2 - A_1} \times (c_2 - c_1) + c_1$$

where

A_0 is the absorbance corresponding to lead or cadmium of the extraction solution;

A_1 is the absorbance corresponding to lead or cadmium of the lower bracketing solution;

A_2 is the absorbance corresponding to lead or cadmium of the upper bracketing solution;

c_1 is the lead or cadmium concentration, expressed in milligrams per litre, of the lower bracketing solution;

c_2 is the lead or cadmium concentration, expressed in milligrams per litre, of the upper bracketing solution.

NOTE — If the extraction solution was diluted, an appropriate correction factor has to be used in the equation.

10.2 Calibration curve technique

Read the lead or cadmium concentration, expressed in milligrams per litre of extraction solution, directly from the calibration curve.

10.3 Calculation of release of lead and cadmium for flatware

The lead or cadmium released per unit surface area from flatware, a_0, expressed in milligrams per square decimetre, is given by the equation

$$a_0 = \frac{c_0 \times V}{S_R}$$

where

c_0 is the lead or cadmium concentration, expressed in milligrams per litre, of the extraction solution, calculated as specified in 10.1 or 10.2;

V is the volume, in litres, of test solution used for the extraction (see 9.1.2);

S_R is the reference surface area (see 8.3.2), expressed in square decimetres, of the test specimen.

10.4 Reporting

For hollow-ware, report the results to the nearest 0,1 mg/l for lead and to the nearest 0,01 mg/l for cadmium.

For flatware, report the results to the nearest 0,1 mg/dm² for lead and to the nearest 0,01 mg/dm² for cadmium.

11 Test report

The test report shall include the following information :

a) a reference to this International Standard;

b) identification of the articles tested, for example whether they were flatware or hollow-ware;

c) the number of specimens tested;

d) each single result, in accordance with 10.4;

e) any unusual features noted during the determination;

f) any operation not included in this International Standard, or regarded as optional;

g) whether each single specimen satisfies the requirements for permissible limits of release as specified in ISO 7086/2.

International Standard **ISO** 7086/2

INTERNATIONAL ORGANIZATION FOR STANDARDIZATION●МЕЖДУНАРОДНАЯ ОРГАНИЗАЦИЯ ПО СТАНДАРТИЗАЦИИ●ORGANISATION INTERNATIONALE DE NORMALISATION

Glassware and glass ceramic ware in contact with food — Release of lead and cadmium — Part 2 : Permissible limits

Articles en verre et en vitro céramique en contact avec les aliments — Émission de plomb et de cadmium — Partie 2 : Limites admissibles

First edition — 1982-11-15

UDC 666.172.3/.5 : 615.9

Ref. No. ISO 7086/2-1982 (E)

Descriptors : glassware, tableware, lead, cadmium, limits.

Foreword

ISO (the International Organization for Standardization) is a worldwide federation of national standards institutes (ISO member bodies). The work of developing International Standards is carried out through ISO technical committees. Every member body interested in a subject for which a technical committee has been set up has the right to be represented on that committee. International organizations, governmental and non-governmental, in liaison with ISO, also take part in the work.

Draft International Standards adopted by the technical committees are circulated to the member bodies for approval before their acceptance as International Standards by the ISO Council.

International Standard ISO 7086/2 was developed by Technical Committee ISO/TC 166, *Ceramic ware, glassware and glass ceramic ware in contact with food*, and was circulated to the member bodies in May 1981.

It has been approved by the member bodies of the following countries :

Austria	Israel	South Africa, Rep. of
Brazil	Japan	Spain
Canada	Mexico	United Kingdom
Egypt, Arab Rep. of	Poland	USA
France	Romania	

The member bodies of the following countries expressed disapproval of the document on technical grounds :

Czechoslovakia
Germany, F. R.

Glassware and glass ceramic ware in contact with food — Release of lead and cadmium — Part 2 : Permissible limits

0 Introduction

The problem of lead and cadmium release from glassware and glass ceramic ware requires effective means of control to ensure the protection of the population against possible hazards arising from the use of improperly formulated, applied or fired glazes and/or decorations on the food contact surfaces of glassware and glass ceramic ware used for the preparation, serving and storage of food and drinks. As a secondary consideration, different requirements from country to country for the control of the release of toxic materials from the surfaces of glassware and glass ceramic ware present non-tariff barriers to international trade in these commodities. Accordingly, there is a need to establish internationally accepted permissible limits for the release of lead and cadmium from glassware and glass ceramic ware.

An expert panel, convened by the World Health Organization (WHO), met in Geneva, in June 1976, and recommended the adoption of sampling methods, testing procedures and limits for the release of toxic materials from ceramic ware.[1] A further meeting was convened by WHO in November 1979.[2]

The permissible limits specified in this International Standard are based on the WHO recommendations, because it was the sense of the WHO meeting that the term "ceramic" includes ceramics, glass, vitreous enamels and glass ceramics. As the capability of the industry increases, efforts will be made to reduce these limits for lead and cadmium release.

The results of an international survey showed that cooking ware made from glass or glass ceramics is not normally decorated on the food contact surfaces. For that reason this International Standard does not address cooking ware.

1 Scope

This part of ISO 7086 specifies permissible limits for the release of lead and cadmium from glassware and glass ceramic ware intended for use in contact with food (including drinks).

2 Field of application

This part of ISO 7086 is applicable to articles made from glass and glass ceramics, which may be transparent, translucent, opaque, coloured, colourless or decorated on the food contact surface, and which are intended to be used for the preparation, serving and storage of food, including packaging.

It does not apply to vitreous and porcelain enamel ware, nor to ceramic ware.

3 Reference

ISO 7086/1, *Glassware and glass ceramic ware in contact with food — Release of lead and cadmium — Part 1 : Method of test.*

4 Definitions

See ISO 7086/1.

5 Permissible limits

The lead and cadmium release from any individual article, when determined by the method specified in ISO 7086/1, shall not exceed the values given in the table.

NOTE — These values are expressed in milligrams per square decimetre of the reference surface area for flatware and in milligrams per litre of extraction solution for hollow-ware.

Table

Type of glassware and glass ceramic ware	Maximum lead release		Maximum cadmium release	
	mg/dm²	mg/l	mg/dm²	mg/l
Flatware	1,7		0,17	
Small hollow-ware		5,0		0,50
Large hollow-ware		2,5		0,25

---

1) See WHO/Food Additives 77.44. *Ceramic Foodware Safety, Sampling, Analysis and Limits for Release* (Report of a WHO meeting, Geneva, 8-10 June 1976).

2) See WHO/Food Additives HCS/79.7. *Ceramic Foodware Safety, Critical Review of Sampling, Analysis and Limits for Lead and Cadmium Release* (Report of a WHO meeting, Geneva, 12-14 November 1979).

14
Phase Diagrams

The following are phase diagrams for lead oxides published in "Phase Diagrams for Ceramists," The American Ceramic Society, Columbus, OH:

1.	$PbO–SiO_2$	Fig. 284
2.	$PbO–SiO_2$	Fig. 4352
3.	$PbO–SiO_2$	Fig. 5170
4.	$PbO–SiO_2$	Fig. 5172
5.	$PbO–SiO_2$	Fig. 5173
6.	$PbO–GeO_2$	Fig. 2329
7.	$PbO–GeO_2$	Fig. 5168
8.	$PbO–B_2O_3$	Fig. 2327
9.	$PbO–B_2O_3$	Fig. 4349
10.	$PbO–B_2O_3$	Fig. 5167
11.	$PbO–Al_2O_3$	Fig. 280
12.	$PbO–Fe_2O_3$	Fig. 282
13.	$PbO–SnO_2$	Fig. 285
14.	$PbO–Nb_2O_5$	Fig. 287
15.	$PbO–P_2O_5$	Fig. 288
16.	$PbO–Ta_2O_5$	Fig. 289
17.	$PbO–V_2O_5$	Fig. 290
18.	$PbO–CrO_3$	Fig. 291
19.	$PbO–MoO_3$	Figs. 292 and 4355
20.	$PbO–Bi_2O_3$	Fig. 326
21.	$PbO–SiO_2–Al_2O_3$ (high Al_2O_3 & SiO_2 portion)	Fig. 396
22.	$PbO–SiO_2–Al_2O_3$ (high PbO portion)	Fig. 397
23.	$PbO–SiO_2–BaO$	Figs. 2460–63
24.	$PbO–SiO_2–B_2O_3$*	Fig. 4585 (LSP#11)
25.	$PbO–SiO_2–B_2O_3$ (above 75% PbO)	Fig. 4584
26.	$PbO–SiO_2–K_2O$	Figs. 174 and 175
27.	$PbO–SiO_2–MgO$	Fig. 2513
28.	$PbO–SiO_2–Na_2O$	Fig. 216
29.	$PbO–B_2O_3–TiO_2$	Fig. 743 (1964)
30.	$PbO–Bi_2O_3–MoO_3$	Figs. 744 (1964) and 2552
31.	$PbO–Bi_2O_3–WO_3$	Fig. 2553 (1969 Supplement)
32.	$PbO–SiO_2–P_2O_5$ (high PbO)	Fig. 746 (1964)
33.	$PbO–TiO_2–ZrO_2$	Fig. 747 (1964)

34. PbO–TiO$_2$–V$_2$O$_5$	Fig. 748 (1964), 2560, and 2561
35. PbO–SrO–SiO$_2$*	(LSP#13)
36. PbO–CaO–SiO$_2$*	(LSP#12)
37. PbO–B$_2$O$_3$–Fe$_2$O$_3$–PbF$_2$–Y$_2$O$_3$	Fig. 6058
38. PbO–KF–Nd$_2$O$_3$–Ta$_2$O$_5$–ZrO$_2$	Fig. 6057

*Available from the American Ceramic Society as large-scale diagrams, 50 cm on the side. Order numbers are given in parentheses.

15
Bibliography

Ceramic Glazes

C. W. Parmelee; Ceramic Glazes. Industrial Publications, Inc., Chicago, 1948.

J. H. Koenig and W. H. Earhart; Literature Abstracts of Ceramic Glazes, College Offset Press, Philadelphia, PA, 1951.

R. F. Geller, E. N. Bunting, and A. S. Creamer, "Some Soft Glazes of Low Thermal Expansion," *Natl. Bur. Stand. J. Res.*, **20**, 57–66 (1938).

H. Harkort, "Lead-Stable Glazes and Their Testing," *Keram. Rundsch.*, **46** [41] 481–84 (1938).

J. H. Koenig, "Lead Frits and Fritted Glazes"; 116 pp. in Engineering Experiment Station Bulletin. The Ohio State University, Columbus, OH, 1937.

R. Rieke and H. Mields, "Acid Resistance of Ceramic Lead Frits in Relation to Composition," *Ber. Dtsch. Keram. Ges.*, **16** [7] 331–49 (1935).

J. W. Mellor, "Durability of Frits, Glazes, Glasses and Enamels in Service," *Trans. Engl. Ceram. Soc.*, **34**, 113 (1934–35).

H. Harkort, "Producing Lead Glazes Non-Injurious to Health of Workers," *Sprechsaal*, **67** [41] 621–23 (1934); **67** [42] 637–39 (1934).

V. V. Vargin and L. L. Fradhova, "Lead Glazes for Faience," *Ceram. Glass*, **7** [5] 29–34 (1934).

D. T. Shaw, "Color Formation in Raw Lead Glazes," *J. Am. Ceram. Soc.*, **15** [1] 36–60 (1932).

Dueffe, "Use of Lead Glaze in Practice," *Ind. Silic.*, **8** [8] 14 (1930).

H. Mohl and H. Lehmann, "Yellow Color of Lead Glazes," *Sprechsaal*, **62** [26] 463–65 (1929).

W. G. Whitford, "Raw Lead Glazes," *Trans. Am. Ceram. Soc.*, **14**, 312–30 (1917).

A. S. Watts, "Lead Oxide versus Lead Carbonate in Glazes," *Trans. Am. Ceram. Soc.*, **17**, 474–83 (1915).

J. W. Mellor, "Use of Lead in Fritted Compounds," *Trans. Engl. Ceram. Soc.*, **12**, 1–18 (1912).

W. P. Rix, "Lead Frits and Their Adaptation to Pottery Glazes," *Trans. Am. Ceram. Soc.*, **4**, 201–7 (1902).

Glass

F. V. Tooley; Handbook of Glass Manufacture. Ogden Publishing Co., New York, 1953.

J. E. Stanworth; Physical Properties of Glass. Oxford University Press, Toronto, 1950.

S. R. Scholes; Modern Glass Practice. Industrial Publications, Inc., Chicago, 1946.

C. J. Phillips; Glass: The Miracle Maker. Pitman, New York, 1941.

G. W. Morey, The Properties of Glass. Reinhold Publishing Corp., New York, 1938.

G. F. Brewster and N. J. Kreidl, "Radiation-Absorbing Glasses," *J. Am. Ceram. Soc.*, **35** [10] 259–64 (1952).

K. Sun, "Fundamental Condition of Glass Formation," *J. Am. Ceram. Soc.*, **30** [9] 277 (1947).

G. J. Bair, "The Constitution of Lead Oxide-Silica Glasses I Atomic Arrangement," *J. Am. Ceram. Soc.*, **19**, 339 (1936).

W. H. Zachariasen, "The Atomic Arrangement in Glass," *J. Am. Chem. Soc.*, **54**, 3841 (1932).

"Low Expansion Lead Borosilicate Glass of High Chemical Durability," U.S. Patent 2 570 020, Oct. 2, 1951.

H. H. Blau, "Silica-Free Glass," (Useful for transmission of rays of relatively long wave lengths), U.S. Patent 2 701 208, Feb. 1, 1955.

Enamels

A. I. Andrews; Enamels, 2nd ed. The Twin City Printing Co., Champaign, IL, 1945.

E. H. McClellan; "Enamel Bibliography and Abstracts" (1928 to 1939). American Ceramic Society, Columbus, OH, 1944.

R. L. Fellows; "Enamel Bibliography and Abstracts" (1940–1949). American Ceramic Society, Columbus, OH, 1953.

L. E. Hansen; Manual of Porcelain Enameling. Ferro Enamel Corp., Cleveland, OH, 1937.

Anon., "Use of Lead in Enamels," *Ceram. Ind.*, **46** [4] 92 (1946).

Anon., "Why Not Lead-Fluxed Porcelain Enamel?," *Ceram. Ind.*, **46** [2] 69–70 (1946).

H. E. Simpson, "Development of an Enamel on a Eutectic Basis," *J. Am. Ceram. Soc.*, **13** [1] 62–79 (1930).

A. Cornille, "Characteristic Tests and Uses of Lead Compounds in Enamels," *Ceram. Verrerie*, **49**, 149–52, 221–25, 283–86, 347–50, 395–98, 445–48 (1929).

H. F. Staley, "Materials and Methods Used in the Manufacture of Enameled Cast Iron Wares," *Natl. Bur. Stand. Tech. Paper* 142, 1919.

J. H. Handwerk and T. N. McVay, "Enamels for Metals," U.S. Atomic Energy Commission, ORO-47, University of Alabama.

School of Architecture, Princeton University, "Curtain Walls of Stainless Steel," American Iron and Steel Institute. (1955).

A. J. Deyrup et al., U.S. Patents 2 316 745 (1943), 2 352 425 (1944), 2 467 114 (1949), 2 544 139 (1951), 2 642 364 (1953), 2 653 877 (1953), assigned to E. I. Du Pont de Nemours & Co., Wilmington, DE.

Paul A. Huppert, "How We Got Where We Are in Light-Metal Enameling," *Ceram. Ind.*, **67** [1] 64–5, 97–98 (1956).

Ferroelectrics and Piezoelectrics

W. P. Mason; Piezoelectric Crystals and Their Application to Ultrasonics.

G. Shirane and S. Hoshino, "X-ray Study of Phase Transitions in $PbZrO_3$ Containing Ba or Sr," *Acta Crystallogr.*, **7** [2] 203–10 (1954).

B. Jaffe et al., "Piezoelectric Properties of Lead Zirconate — Lead Titanate Solid Solution Ceramics," *J. Appl. Phys.*, **25** [6] 809–10 (1954).

G. Goodman, "Ferroelectric Properties of Lead Metaniobate," *J. Am. Ceram. Soc.*, **36** [11] 368–72 (1953).

A. P. de Bretteville, Jr., "Antiferroelectric $PbZrO_3$ and ferroelectric $BaTiO_3$ Phenomena," *Ceram. Age*, **64** [4] 18–22, 92–93 (1953).

G. Shirane and S. Hoshino, "Crystal Structure of the Ferroelectric Phase in $PbZrO_3$ Containing Ba or Ti," *Phys. Rev.*, **86** [2] 248 (1952).

G. Shirane et al., "Phase Transition in Lead Zirconate," *Phys. Rev.*, **80** [3] 485 (1950).

S. Roberts, "Dielectric Properties of Lead Zirconate and Barium Lead Zirconate," *J. Am. Ceram. Soc.*, **33** [2] 63–66 (1950).

"Piezoelectric Transducers Using Lead Titanate and Lead Zirconate," U.S. Patent 2 708 243, May 10, 1955.

Hygiene

"The Control of the Lead Hazard in the Storage Battery Industry," Investigation Report No. 6 and Public Health Bulletin 262, State of California, Department of Public Health, Bureau of Adult Health.

National Safety Council, Safety Reprint 60, National Safety News, August, 1954.

"Occupational Lead Exposure and Lead Poisoning," American Public Health Association, 1943.

W. Schweisheimer, "Lead Exposure in the Glass Industry," *Am. Glass Rev.*, April, 1955.

"Removal of Dust, Gases and Fumes," The Industrial Code, Rule No. 12, State of New York, Department of Labor, Board of Standards and Appeals.

W. Schweisheimer, "The Ceramic Industry and Health," *Am. Glass Rev.*, September, 1955.

Miscellaneous

J. G. N. Braithwaite, "Infrared Filters Using Evaporated Layers of Lead Sulfide, Lead Selenide, and Lead Telluride," *J. Sci. Instrum.*, **32** [1] 10–11 (1955).

"Composite Lead Chromate — Lead Silicate Pigment," U.S. Patent 2 668 122, Feb. 2, 1954.

"Composite Lead Sulfate — Lead Silicate Pigment," U.S. Patent 2 477 277, July 26, 1946.

E. M. Levin, H. F. McMurdie, and F. P. Hall; Phase Diagrams for Ceramists. American Ceramic Society, Columbus, OH, 1956; see also Vols. 2–5 of Phase Diagrams for Ceramists.

J. E. Ablard, "Strains in Glass Produced by Applied Color Labels," *J. Am. Ceram. Soc.*, **28** [7] 189–95 (1945).

16
Epilogue

Lead is a useful element that has served mankind well for centuries beyond count. Lead and its compounds are versatile materials performing a great variety of essential functions better than any other available material or combination of materials. Yet lead is under attack by well-meaning, but perhaps insufficiently informed, regulators, environmentalists, and trade negotiators, whose actions could deprive us of this essential commodity. What would be gained if lead were denied us? What would be the penalties if such an event should occur? The following is a brief review which, it is hoped, will help to answer these questions.

Sometime before 6500 B.C., early man began to learn how to produce and use lead. It is one of the oldest metals known to man, and its discovery is lost in antiquity. Lead was in common use in biblical times and is mentioned in the Old Testament. The ancients learned how to form sheet lead and roll sheet into pipe. They mastered the arts of lead burning and soldering. Sheet lead was used to line the Hanging Gardens of Babylon in order to retain the moisture needed by the vegetation. Lead was also used for other purposes in construction that are still common practice today. The sign of Saturn, ♄, father of the gods, was used as the symbol for lead by the ancients as well as by later alchemists, whose main interest seems to have been that of trying to convert lead into gold. With today's inflation, we may yet succeed.

The ancient Romans called it *plumbum*. That is a good Latin word still used in various forms in the English language. "Plumb," "plumbing," "plumber," "plumb line," and "plumb bob" all derive from the Latin original. The Romans used lead in many ways. They made great quantities of lead pipe for transporting water. Some of this ancient pipe is still to be seen in Rome, Pompeii, and Herculaneum in Italy as well as at Bath in England. Some is still in good working condition. Women used red lead as a cosmetic, and lead oxide was an important constituent of glass and pottery glazes. This is the forerunner of wide use in today's vast assortment of ceramic products.

Lead and its compounds are indispensable in today's economy. It is probably not an overstatement to say that lead is essential to civilization as we know it. Without lead, our mechanized transport system would be in serious difficulty. Certainly there would be few, if any, internal combustion engines. Automobiles, locomotives, airplanes, trucks, buses, golf carts, snowmobiles, power boats, etc. could not function acceptably without storage batteries, to mention only one of many uses for lead or its compounds in our transportation equipment.

Without lead glass, modern lighting would be greatly limited and there would probably be no television. Many vital electronic systems depend on lead oxides for their very existence. Much optical glass and ophthalmic glass would not exist, along with many other glass products which we take for granted.

The principal ore from which primary lead metal is smelted is galena, a crystalline form of lead sulfide (PbS). This interesting material is a semiconductor. Many of our senior citizens will remember the crystal radio sets of earlier days. Those little gray crystals that they tickled with the "cat's whisker" were pieces of galena. Galena is roasted, smelted, and refined to recover pure lead metal. Frequently, silver and zinc accompany lead in its ores and must be separated. In fact, lead ore is one of the principal sources of new silver.

Soluble compounds of lead can be hazardous to those exposed, if not handled carefully. The solubility of galena in gastric and other body juices is so low that it presents little health hazard. Lead poisoning is not a significant hazard in the lead mining industry.

Modern transport could not exist without the lead–acid storage battery. Why would not some other type of battery, such as the nickel–cadmium battery, do as well? For one thing, there simply is not enough of either metal to fill the needs. There is also the matter of economics. The cost of a nickel–cadmium battery would be more than 10 times that of a comparable lead–acid battery. Another compelling factor in favor of the lead–acid battery is its ability to deliver energy at low temperatures. Most other systems, proposed or under development, do not perform well at 0°F or below. In addition, most of the lead used in storage batteries is recovered and recycled back into the supply system. This again is a powerful economic plus in favor of the lead–acid battery. Finally, they are here now, and available as needed, while continuing engineering advances are providing greatly improved power to weight ratios.

There are many types of lead–acid storage batteries manufactured for a variety of purposes. Original equipment batteries for new vehicles represent a market of 8–12 million per year in the U.S. alone. Much greater than this, however, is the U.S. market for replacement batteries for vehicles already in service. This market has reached close to 60 million batteries per year. We also have the large industrial battery industry, whose batteries of specific design serve many specialized needs. Special batteries are made for hospitals, telephone exchanges, computers, and other emergency standby power purposes so that essential services may continue in the event of a power failure. Others are made for earth-moving equipment, submarines, and industrial equipment including lift trucks. At the other end of the size spectrum are special industrial batteries of the lead–acid type used in an assortment of power hand tools and electronic toys. One unusual type is a small battery of cylindrical design used in miners' safety lamps. Again, in the very large category now beginning to appear are banks of large battery units used for load leveling at power stations. They store energy during the low-demand period of the day and deliver it during the peak-load period. Lead–acid batteries of advanced design and engineering have and will continue to have a very important place in electric vehicles. These nonpolluters must play a large and growing part in our transport system, especially small automobiles, light trucks, and vans.

Storage battery manufacturers are the largest users of lead, representing over 65% of the total. It is important to note that while this lead is *used,* it is *not all consumed*. Most of the lead used to make storage batteries is eventually recycled. It is smelted and refined as secondary lead (as distinguished from primary metal made from ore). Somewhat less than half of the lead content of a battery is represented by the metal parts, the grids, connectors, and terminals. Most are made from secondary lead and some from primary lead. The balance, the active material in the battery plates, begins as lead oxide. Much of this is made from primary or virgin lead. Thus, with each cycle of recovery of battery scrap, the amount of production of secondary lead grows. *Lead is unique in that it is the only essential metal whose supply increases the more it is used.* In 1979, about 800 000 tons of secondary lead was produced in the United States, up from 240 000 tons in 1930. The tonnage of primary lead produced from domestic ore and some imported ore was about 650 000 tons in 1979.

A growing use for lead sheet and lead powder mixed in vinyl sheet or certain foamed plastics is for sound attenuation, or noise control. Properly used, these materials greatly reduce the sound level and protect the hearing of workers in the vicinity of loud machinery and of passengers inside jet airplanes. Lead metal also serves well as a protective covering for underground and submarine cables used for communication.

Lead plays a most important part in the plastics industry. Poly(vinyl chloride) (PVC) is one of the most versatile and widely used of the plastics and it is used extensively in the electric wire covering industries. PVC's must be stabilized to prevent degradation of electrical and physical properties by heat and light. Of the known vinyl stabilizers, only compounds of mercury, silver, and lead form chloride salts which are essentially insoluble in water. Therefore, mercury, silver, or lead stabilizers are the most suitable for electric wire and cable covering and other electronic wire covering uses. Since mercury is more toxic and more costly than lead and the cost of silver is exorbitant, lead stabilizers are the most practical for the important use of vinyl wire insulation. This means that automotive ignition and wiring systems and aircraft, housing, and industrial wiring systems depend upon lead stabilizers to produce a satisfactory wire covering. There are many other valuable uses for lead-stabilized vinyl plastics, among which is vinyl sewage effluent piping and fittings.

Lead in a variety of forms is used in many ways to protect mankind against radiation. Aprons containing lead powder in vinyl or rubber sheet protect X-ray technicians as well as the patients. Lead glass, some containing more than 80% lead oxide, is used in viewing ports and windows in nuclear installations where radiation is intense. Lead metal in brick, slab, or sheet form is used extensively as well, while lead canisters are increasingly important in transporting and storing radioactive materials. Lead glass containing 30–35% lead oxide is used in the manufacture of TV picture tubes, both color and black and white, to absorb and block gamma radiation from the electron gun. We depend on lead for radiation protection because it is the most effective and versatile material for the purpose, it is readily available in the quantity needed, and it is less costly than gold, which could be used, although in metallic form only. *Lead serves us well in protecting against radiation.*

A substantial segment of the lead industry is represented by the manufacturers of lead oxides, lead silicates, and certain of the lead pigments. These chemicals, because of an assortment of properties not found in any other material or combination of materials, fill a vast number of essential uses in many industries. Red lead is an important anticorrosive pigment used in paints which protect our bridges, ship hulls, and other steel structures from rusting. Lead chromate is used also as a rust-inhibitive pigment, and because of its highly visible bright yellow color, it is used in paints for traffic signs and for traffic marking, among others.

Lead oxide is a powerful flux or solvent for silica and other constituents of glass, glazes, or vitreous enamels. It permits low-temperature melting of these glassy materials. Because of their low melting and maturing temperatures, lead glazes permit the use of a much broader color pallet than leadless glazes. Many of the coloring compounds would be destroyed by the higher firing temperatures required with leadless glazes. Lead glasses have a wide assortment of valuable qualitites not found in other types. Aside from their generally low melting and forming temperatures, lead glasses have a wide working temperature range. Most lead glass compositions have very high surface and volume electrical resistivity, making them especially valuable in electrical and electronic uses. Lead glasses have high index of refraction and, thus, have a level of brilliance not possessed by most other glasses. One will see this in fine lead crystal. Lead glass used in television picture tubes is opaque to X rays, thus providing necessary protection for viewers and even more so for repairmen and for quality-control personnel in TV assembly plants, who may be working with many sets operating. Color TV picture tubes are assembled after the essential phosphors and the screen are in place. The panel and the funnel are then joined by the use of solder glass. There are several solder glass compositions for a variety of applications besides assembly of color TV picture tubes. Many have high content of lead oxide, some greater than 60%. Without it, they simply would not work. Lead oxides of very high purity are essential constituents of many optical and ophthalmic glass compositions. Optical glass for lenses for such items as microscopes, telescopes, periscopes, binoculars, and cameras and for use in assorted fiber optic equipment such as endoscopes and proctoscopes relies on very pure lead oxides for the required optical properties. The same is true for ophthalmic glass used in eye glasses. *Clear vision depends substantially on lead.*

Lead alloys, including lead-bearing metals and solders, constitute a substantial portion of lead use and consumption. Lead, alloyed with antimony, tin, silver, or other metals in a wide variety of solder compositions, has many essential uses in many industries. It is used to join metals, whether sheet, pipe, or plumbing fittings. Auto body solder is used to smooth the contours, while automobile and air-conditioning radiators are soldered to provide a leak-proof assembly. Lead-based solder is used in substantial quantity by the electronics industry in assembly of circuits. In 1979, about 49 000 tons of special solder was used in electronics. Lead alloys are used also as coating materials for iron and steel and are applied by either hot dipping or electroplating. Lead alloys and solders have many uses based on valuable common properties. Most solders melt at low temperatures, some at less than 200°F. All are durable and, when properly used, provide a permanent, electrically conducting bond.

Defense of our country depends on a great variety of military equipment aside from the people trained to use it. We must have missiles, nuclear submarines, airplanes, tanks, guns, and ammunition. We must have transport for people and equipment. *Lead in a variety of forms plays an important role in all of these.* We must have, in addition to our most sophisticated weapons, the ability to communicate. Some of these communication requirements are little known to the public. For example, in order to keep track of submarines, theirs and ours, we have a vast array of sonar buoys, picking up and transmitting signals. This is vital in anti-submarine warfare. The heart of sonar is an item called a transducer. The best of these are made of a ceramic composition called lead zirconate titanate, usually with doping of certain trace elements. Variations of these compositions are used in stereo record players to convert the vibrations of the stylus to an electric current and then to sound. There are many uses for the transducer.

In all of the many centuries in which lead glazes have been used on dinnerware and on floor and wall tile, no substitute has been found that will provide so many desirable or essential qualities. Lead glazes, properly formulated, properly applied, and properly fired, are among the most resistant to the effects of food acids. They are also the *most* resistant to the destructive effects of strong dishwasher detergents. Most bathtubs are coated with enamels containing a high percentage of lead oxide in chemical combination with the other constituents. Other types of sanitary ware also are coated with lead-based glazes or enamels.

Through the thousands of years during which lead has been used by the human family, it is probably safe to say that it is used only when no acceptable substitute exists and never simply for the love of it. Lead is heavy and sometimes that can be a problem, although for some uses, it is a virtue. Lead and its compounds are moderately expensive. Costs might be cut if cheaper substitutes could be found. Lead compounds, if soluble in stomach and other body acids, are toxic and, if not used with proper protection, care, and concern, will be hazardous to those who may be exposed. These facts are well-known to the producers and to the processors, and proper safety measures are well-known. They are also in common and effective use. Lead and its compounds, in spite of these disadvantages (which incidentally are common to the other heavy metals and their compounds) continue to play not only a useful, but a vital role in world commerce and industry and in our daily life.

Lead indeed is a precious metal, not in cost, but in real value. A strong and viable lead industry is vital to the strength of our economy and our nation. Means must be found to forestall misguided legislation, excessive and punitive regulation, and unwise tariff consessions. These together pose a serious threat to the survival of the lead industry, along with those other industries and jobs that depend upon it. What a grim price we would pay for such a loss! And for what gain?

Index

Lead
in the World of
Ceramics
Addenda

John S. Nordyke

Vello tube drawing (courtesy Corning Glass Works).

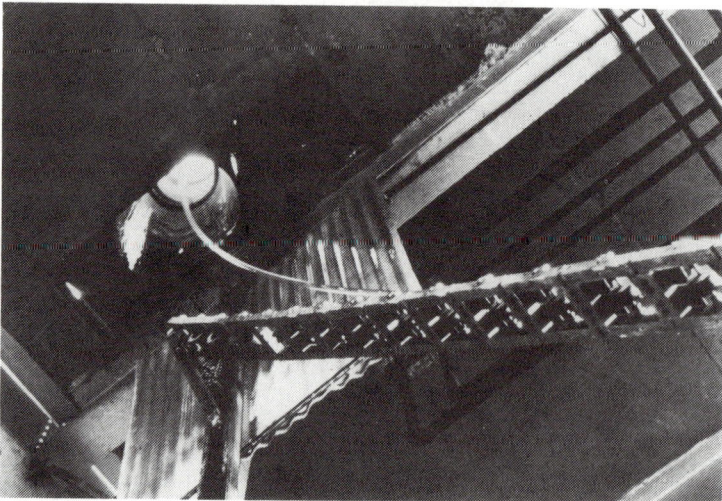

Vello tube drawing (courtesy Corning Glass Works).

ERRATA

Note of Appreciation and **Table of Contents**
p. vii, line 21 and p. xii, line 12: "Mykro/Mycalex" should read "Mykroy/Mycalex."

Chapter 3
pp. 14 and 15: The captions for Figures 1 and 2 should be reversed.
p. 51: Table I, "Approximate Compositions of Solder Glasses," should be in the section on Solder Glasses, beginning on p. 46.

Chapter 4
p. 73: Line 4, first column, should read BeO.

Chapter 10
p. 131, n: "E. I. du Pont de Nemours & Co., Wilmington, DE" should read "Corning Glass Works, Corning, NY."

Chapter 13
p. 167, line 19: "DIS6486" should read "DIS7086."
p. 167, Acknowledgments, line 1: "legend" should read "legion."

Chapter 14
Please note corrected phase diagrams 396, 397 (correct numbers 737, 739), 174, 175 (correct numbers 404, 405). Phase diagrams 24, 35, and 36 are four-color, large-scale diagrams, 50 cm on the side, and are available from The American Ceramic Society at $5 each.

The enclosed photograph shows the Service Plate and Dinner Plate from the White House Service made by Lenox China for President Ronald Reagan. Lead-bearing frits are utilized in achieving the most brilliant and highly reflective glaze surfaces, such as these.

The photograph has an adhesive backing and may be inserted into *Lead in the World of Ceramics* as a frontispiece by removing the paper strip and attaching the photograph to the page opposite the title page.

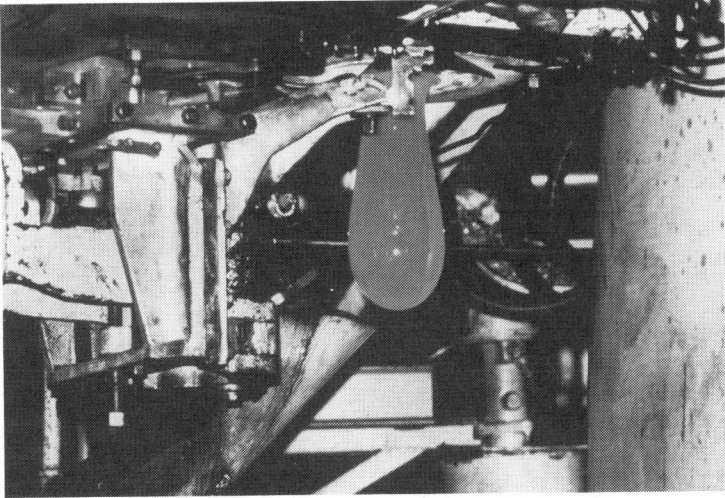

Gob and shear in hot end (courtesy Corning Glass Works).

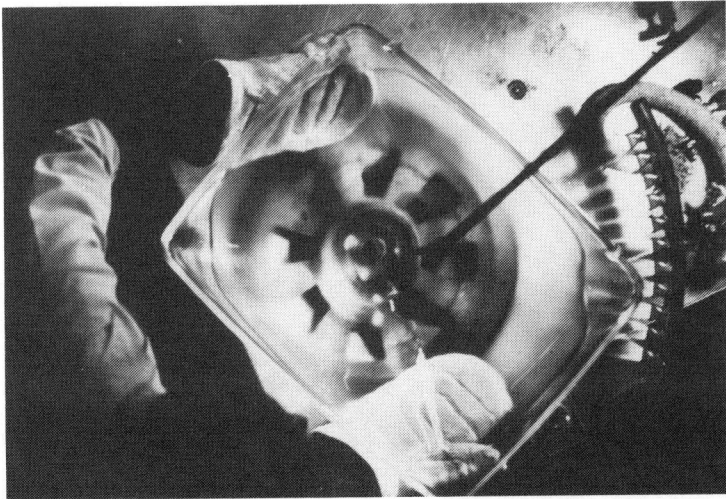

TV face funnel coming out of mold (courtesy Corning Glass Works).

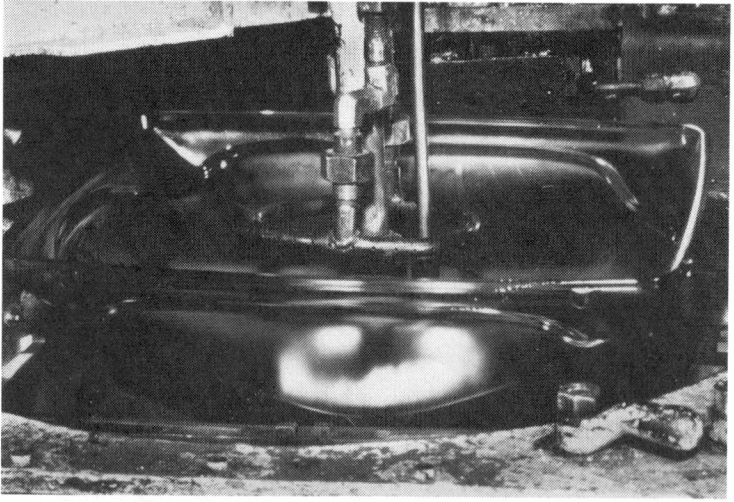

TV face funnel coming out of mold (courtesy Corning Glass Works).

Ophthalmic blanks going into lehr (courtesy Corning
Glass Works).

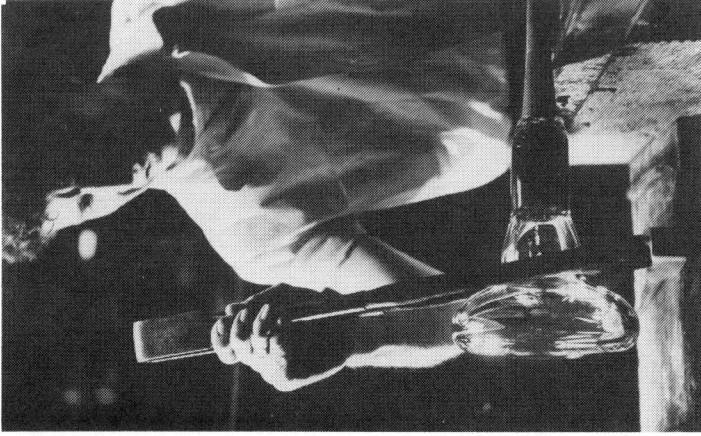

Forming of Steuben glass-initial stage (courtesy Corning Glass Works).

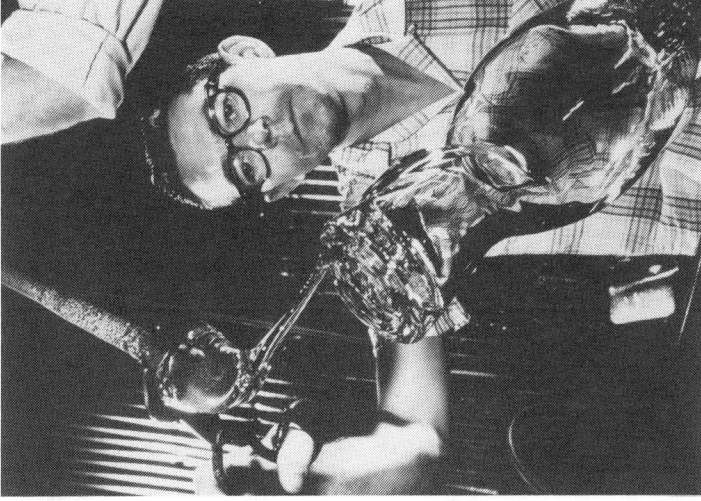

Forming of Steuben glass-intermediate stage (courtesy Corning Glass Works).

Forming of Steuben glass-final stage (courtesy Corning Glass Works).

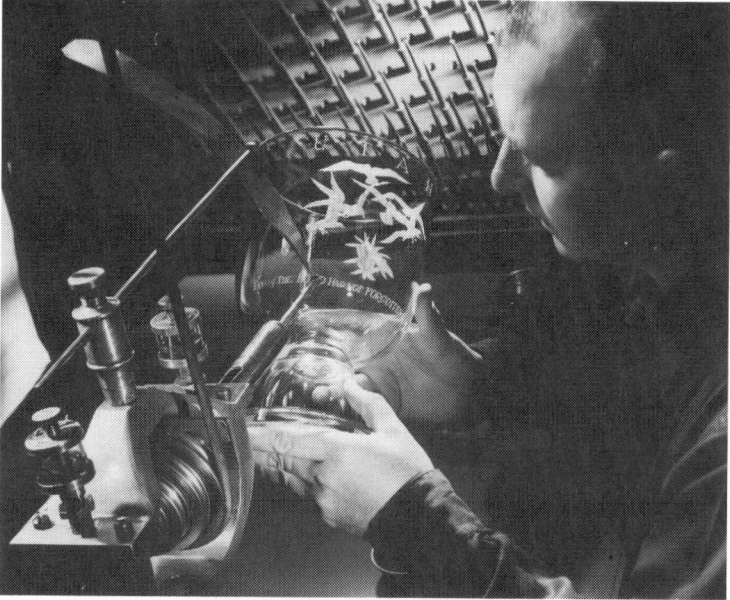

Engraving Steuben glass (courtesy Steuben Glass, Inc.).

Spray glazing line (courtesy Netsch Inc.).

Washington's Dulles International Airport, one of the late Eero Saarinen's last great designs, utilized more than 22 000 linear feet of black porcelain-on-aluminum extrusion for window and door framing, as well as 114 000 square feet of P/E aluminum formed panels.

L. M. Berry Co., Brookfield, Wisconsin.

Northwest Mental Health Center, Columbus, Ohio.

Country Forge heavy-gauge aluminum cookware with SilverStone nonstick surface. The ultraweight aluminum provides fast, even heating. The DuPont premium SilverStone nonstick surface is for stick-free cooking and easy cleaning.

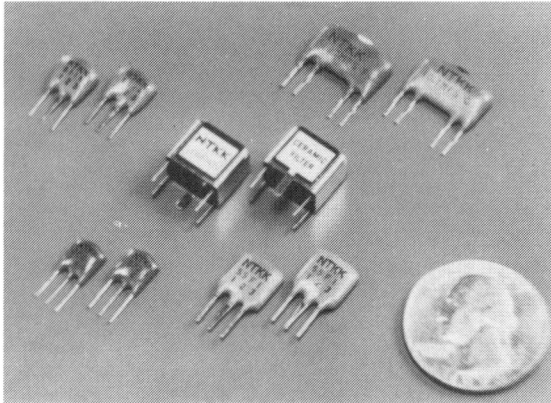

NGK SPARK PLUG CO., LTD.
NAGOYA, JAPAN.

Ceramic Filters

NTKK Ceramic Filters, well associated with the resonant phenomenon of piezoelectric ceramics, have been favorably accepted in various fields of industry. Various intermediate 455 kllz band pass filters for radio, transceiver, pocket-bell, moving and stationary type communication equipments, FM intermediate band pass filter for radio, stereo tuner, sound intermediate band pass filter of TV set, carrier wave filter CB transceiver.

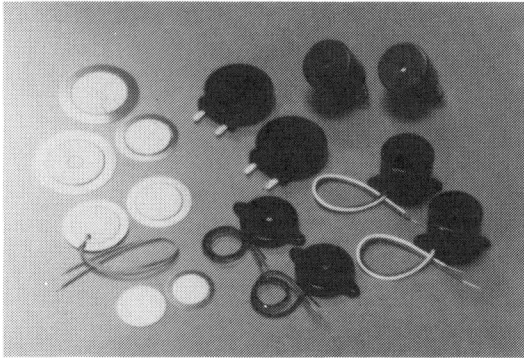

NGK SPARK PLUG CO., LTD.

NAGOYA, JAPAN.

Audio tone transducers

Piezoceramic audio tone transducers produce audible sounds by the vibration of a unit consisting of an integral bonding of the element and a vibration diaphragm.

In comparison with the conventional electromagnetic buzzer, the piezoceramic audio tone transducers have such features as; the current consumption is less than 1/5 of that in magnetic transducers, the construction can be made flat, and the life is semi-permanent because construction has no wearing parts.

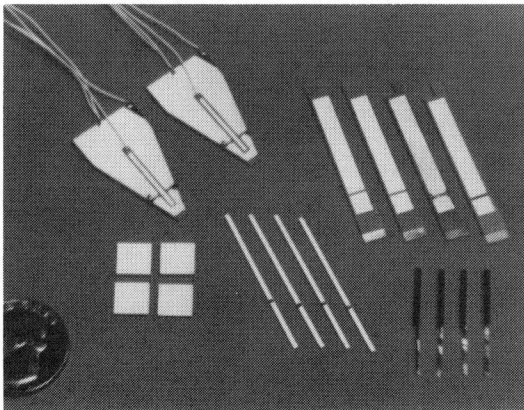

NGK SPARK PLUG CO., LTD.

NAGOYA, JAPAN.

Ceramic Bimorphs

A ceramic bimorph is a high-compliance, high-output, voltage transducer consiting of two thin piezoelectric ceramic plates bonded together. It is used as the pick-up element of medium- and super-sensitive-class stereo set cartridges, other vibration pick-up elements, and transmitting and receiving elements of ultrasonic waves by utilizing its resonant characteristic.

Moreover these bimorphs are utilized as bender elements which are used for ink jet printer, head controller of video tape recorder, actuator for electronic braille point, etc.

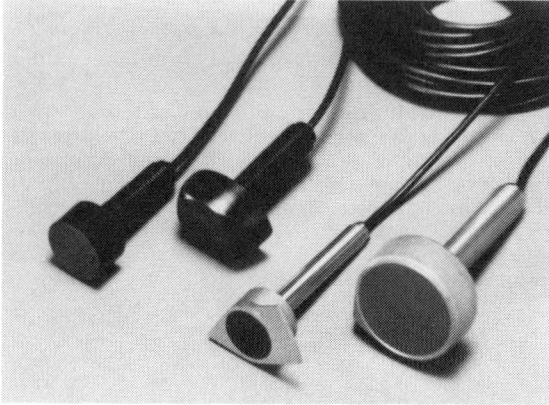

NGK SPARK PLUG CO., LTD.
NAGOYA, JAPAN.

The NTK mold type underwater tranducer is designed to withstand severe conditions in submerged uses, such as, excellent water proof properties, transmission loss, adhesive strength, mechanical strength, thermal variation, etc.

Typical examples for communicative application of these transducers are the fish-finder and the depth-finder in which ultrasonic waves are used as signals.

Recently we developed new underwater transducer which is made by bolt clamped type construction, and it is submerged at bottom of sea for drift-net fishing.

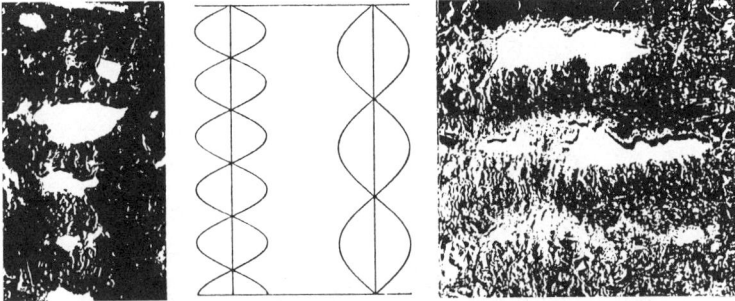

NGK SPARK PLUG CO., LTD.
NAGOYA, JAPAN.

Immersible Switching Ultrasonic Transducer
with two resonant frequency

The newly developed bolt clamped Langevin type transducer has two resonant frequencies. For example, about 25kHz ($\lambda/2$ mode) and 46kHz (λ mode). These transducers are attached into immersible case, thus immersible switching ultrasonic transducer is produced.

This transducer reduces the influence of standing wave. In case of regular immersible transducer, the distance between loops in vibration is about 31mm at 24kHz.

In the immersible switching ultrasonic transducer, (from $\lambda/2$ mode to λ mode, alternately) the position of nodes of $\lambda/2$ mode vibration are covered by the position of loopes of λ mode vibration, therefore uniform sound field is obtained.

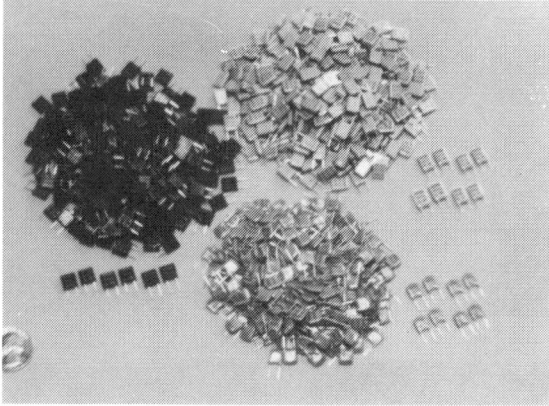

NGK SPARK PLUG CO., LTD.
NAGOYA, JAPAN.

NTKK Ceramic Resonator

NTKK Ceramic Resonator utilizes the resonant characteristic of high-stability piezoelectric ceramics and is suitable for clock oscillator of micro processors. Control and operation functions are the main points for the recent applied fields such as electric equipment, automobile, telephone, communication equipment, camera, duplicator, voice synthesizer and toy. To obtain high-stability of clock frequencies, NTKK Ceramic Resonators are commonly utilized as the optimum resonators.

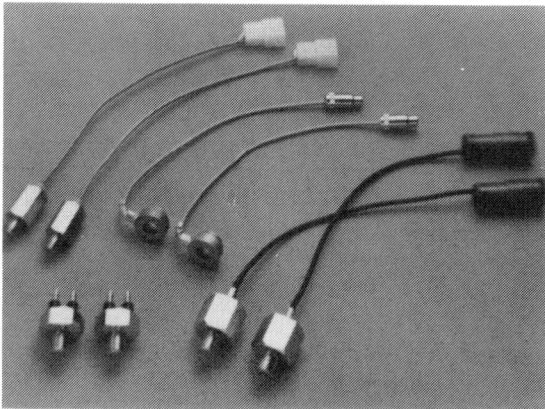

NGK SPARK PLUG CO., LTD.
NAGOYA, JAPAN.

Knock Sensor

We are developing knock sensors using piezoelectric ceramics as an element of sensors, and cn now design knock sensors to meet the customer needs. There are three types of NTK knock sensors, for example, resonant type, nonresonant type and pressure type. The pressure type knock sensor has the characteristic of a good S/N ratio at a high engine speed, and will be installed at the tention bolt in a engine cylinder head.

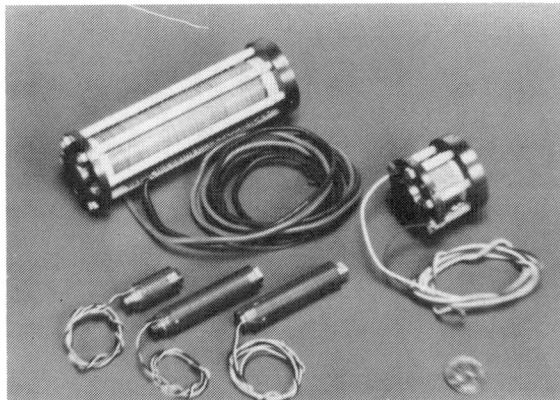

NGK SPARK PLUG CO., LTD.
NAGOYA, JAPAN.

Piezo Stack

NTK Piezo Stack is very accurate positioning actuator. An electrical voltage applied on the piezoelectric material will cause the change in linear dimension of them. One of the advantages of the movement produced by this effect is rapidity with a high accurate positioning without any mechanical conversion.

NTK Piezo Stack is made up of a number of disks which are mechanically arranged in series connection, but electrically drived in parallel connection. As a result, displacement can be enhanced to any desired level by stacking up a multiple number of elements.

The characteristics of the NTK Piezo Stack are accuracy, quick response, fast generated force and long life in positioning. NTK Piezo Stack is, therefore, an ideal actuator for positioning and other various uses.

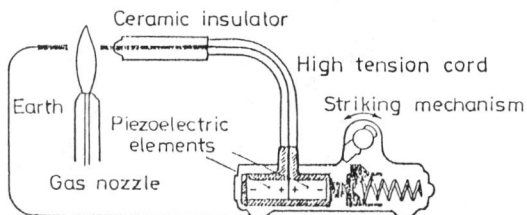

Ceramic insulator · High tension cord · Striking mechanism · Earth · Piezoelectric elements · Gas nozzle

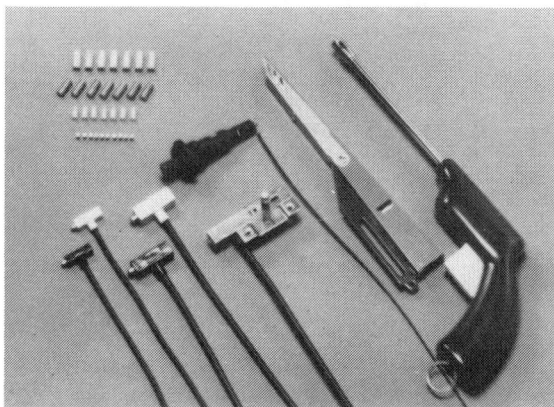

NGK SPARK PLUG CO., LTD.
NAGOYA, JAPAN.

Piezoelectric Igniter

NTK piezoelectric igniter ignites all kinds of gases such as; city gas, propane gas, butane gas, natural gas, etc.

As shown below, the striking mechanism causes the spring loaded striker to hit the convex shaped board of the piezoelectric elements. At the tip of the high tension cord, the electrode (heat resistant nickel wire) covered with a ceramic insulator is placed on the opposite side of the grounding terminal leading from the body of the striking mechanism forming a spark gap. When an electric discharge occurs in a suitable position around the gas nozzle, a high temperature of thousands of degrees is generated by the energy of the spark discharge and thereby the gas is ignited without failure. NTK piezoelectric igniters are used for the gas ranges, gas stoves and various handy lighters.

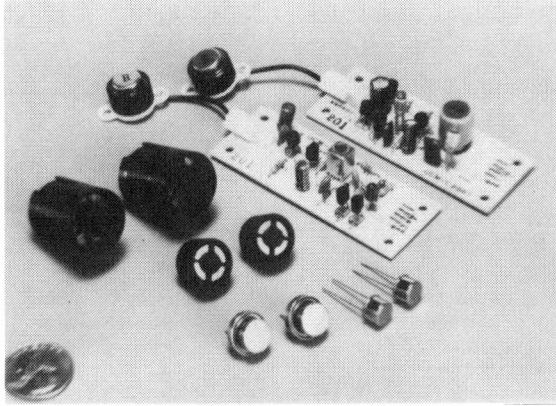

NGK SPARK PLUG CO., LTD.
NAGOYA, JAPAN.

Ultrasonic sensor

Ultrasonic sensor has an enclosed structure of PZT ceramics encapsulated by metal cap, and has features of high temperature stability, small size and light weight.

It is utilized as a pair of transmiter and receiver for remote controller, garage opener, several kinds of sensor for robot, powder detector, liquid level detector, etc.

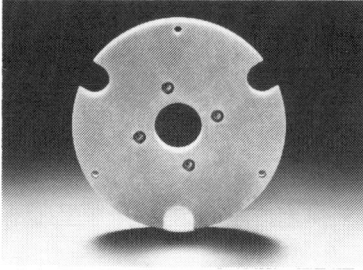

Machined glass-bonded mica disk for high-frequency application with contacts soldered onto silver metallized areas (courtesy Mykroy/Mycalex Co.).

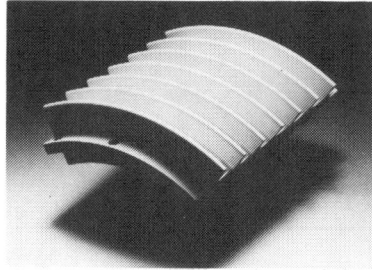

Molded segment of a round commutator for a high-temperature motor; metal rings are placed between barriers (courtesy Mykroy/Mycalex Co.).

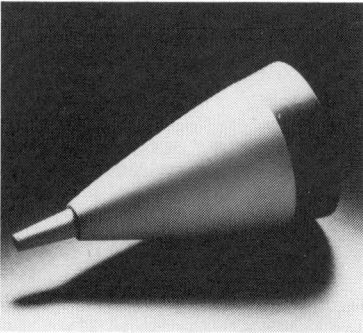

Molded glass-bonded mica radome nose cover for high-frequency radar application; outside edge is metallized for easy assembly (courtesy Mykroy/Mycalex Co.).

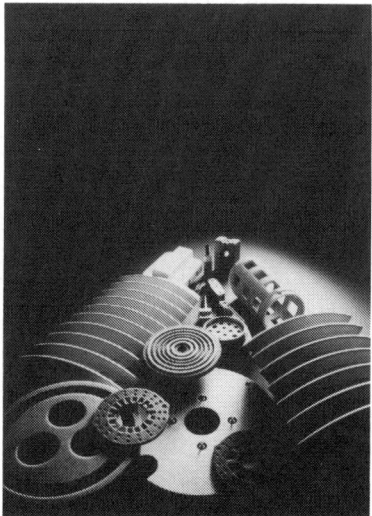

Molded glass-bonded mica parts, some with metal inserts molded in (courtesy Mykroy/Mycalex Co.).

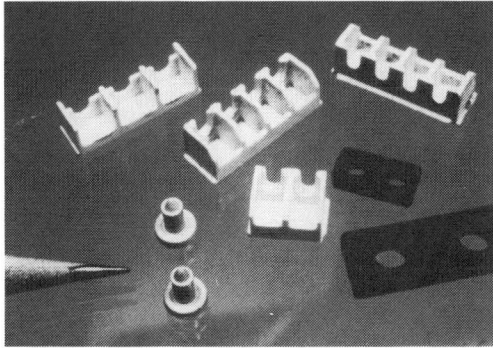

Small and thin-walled (0.025 in.-thick) molded
glass-bonded mica parts (Mykroy/Mycalex Co.).

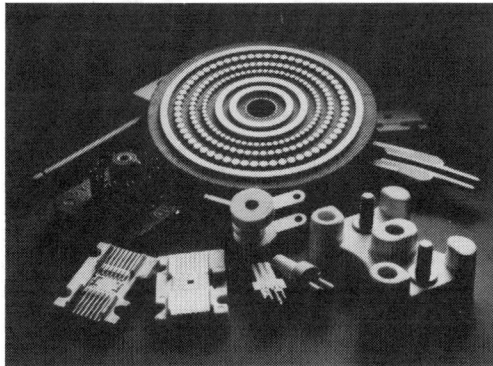

Array of molded glass-bonded mica parts with
metal inserts molded in (courtesy
Mykroy/Mycalex Co.).

PbO–SiO$_2$

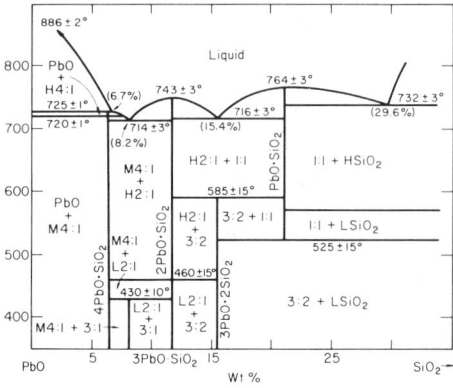

Fig. 4352

PbO–SiO$_2$

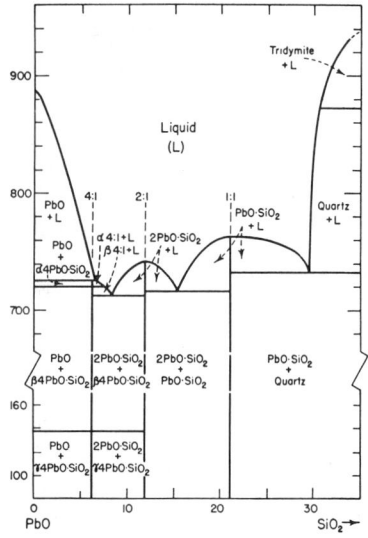

Fig. 284.

PbO–SiO$_2$

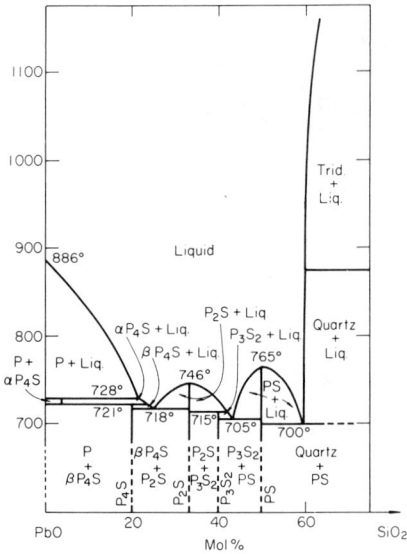

Fig. 5170

PbO–SiO$_2$ (cont.)

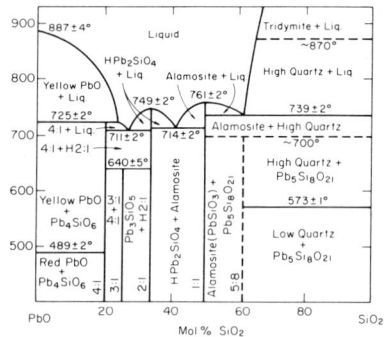

Fig. 5172

PbO–SiO₂

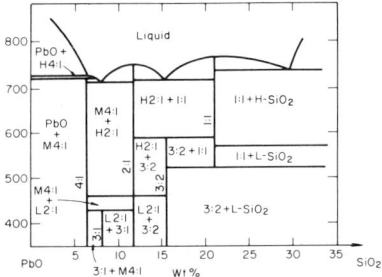

Fig. 5173

PbO–GeO₂

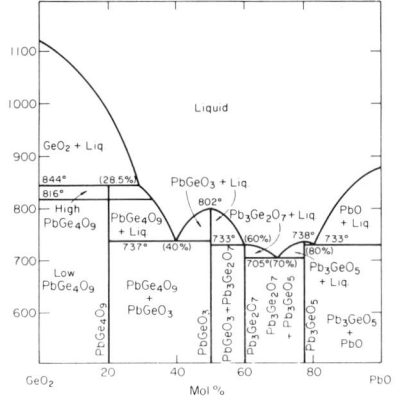

Fig. 5168

PbO–GeO₂

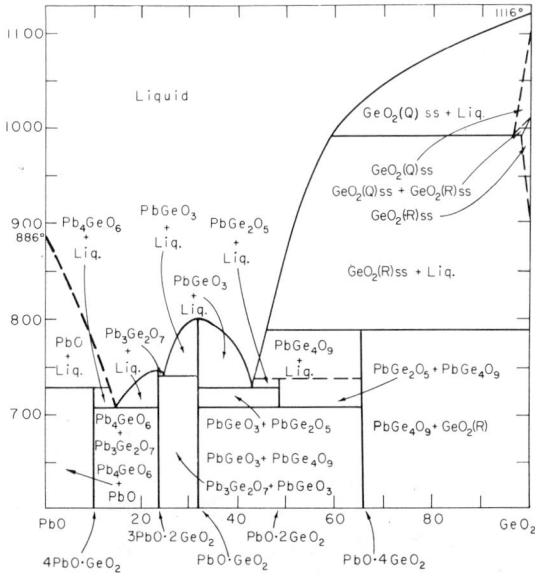

Fig. 2329

PbO–B₂O₃

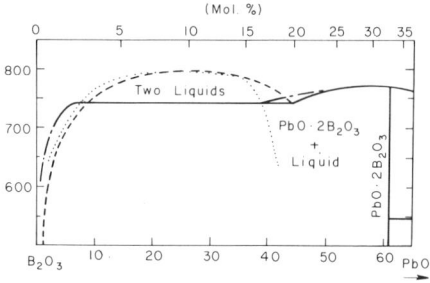

FIG. 2327.—System PbO–B₂O₃ in immiscibility region. Dashed and solid lines represent present diagram of Geller and Bunting; dashed-dot line is proposed liquidus; dotted line represents proposed, two-liquid separation.

PbO–B₂O₃

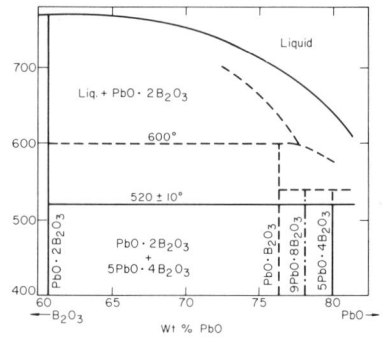

FIG. 4349.—System PbO·2B₂O₃–5PbO·4B₂O₃, revised, showing 3 metastable compounds.

PbO–B₂O₃

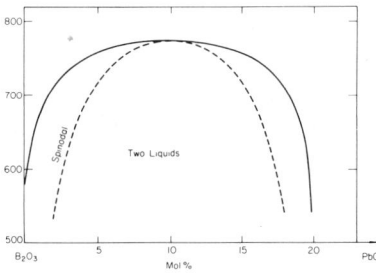

FIG. 5167

PbO–Al₂O₃

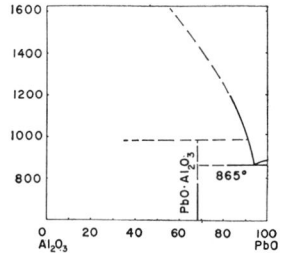

FIG. 280

PbO–Fe₂O₃

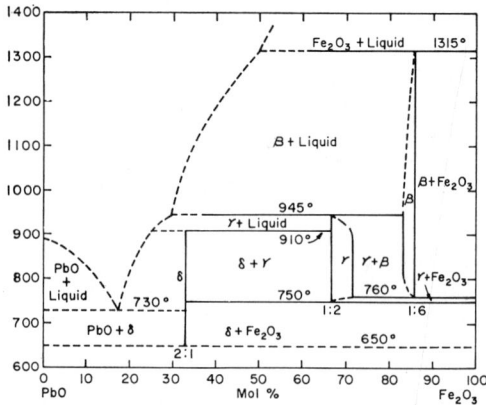

FIG. 282

PbO–SnO₂

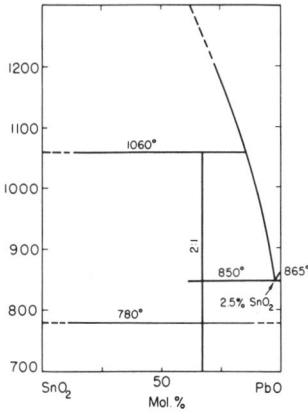

FIG. 285

PbO–Nb₂O₅

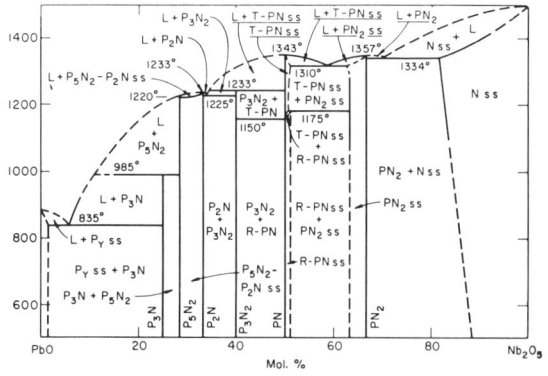

FIG. 287.—System PbO–Nb₂O₅. P_Y—yellow PbO, orthorhombic; P₃N—3PbO·Nb₂O₅; P₄N₂—5PbO·2Nb₂O₅; P₂N—2PbO·Nb₂O₅; P₁N₂—3PbO·2Nb₂O₅; T-PN—tetragonal PbO·Nb₂O₅; R-PN—rhombohedral PbO·Nb₂O₅; PN₂—PbO·2Nb₂O₅; N—Nb₂O₅; ss—solid solution; L—liquid.

PbO–P₂O₅

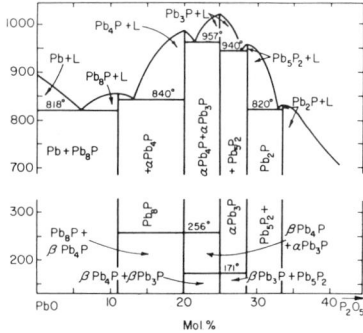

FIG. 288.—System PbO–P₂O₅. Pb = PbO; P = P₂O₅; L = liquid.

PbO–Ta₂O₅

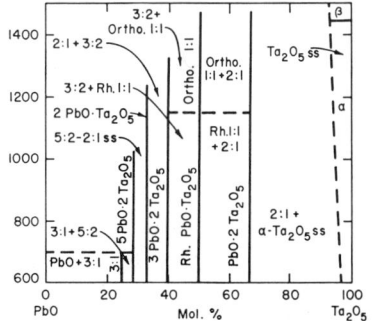

FIG. 289.—System PbO–Ta₂O₅; subsolidus. Ortho = orthorhombic modification; Rh = rhombohedral modification. Compounds in two-phase areas designated as mole ratios of components.

E. C. Subbarao, *J. Am. Ceram. Soc.*, **44** [2] 93 (1961).

PbO–V₂O₅

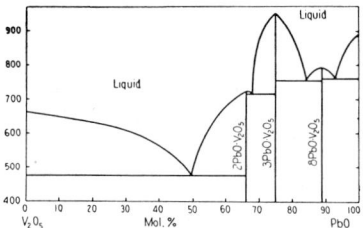

FIG. 290

PbO–CrO₃

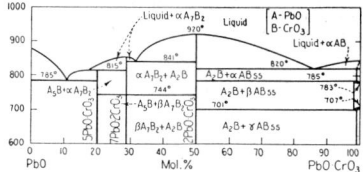

FIG. 291

PbO–MoO₃

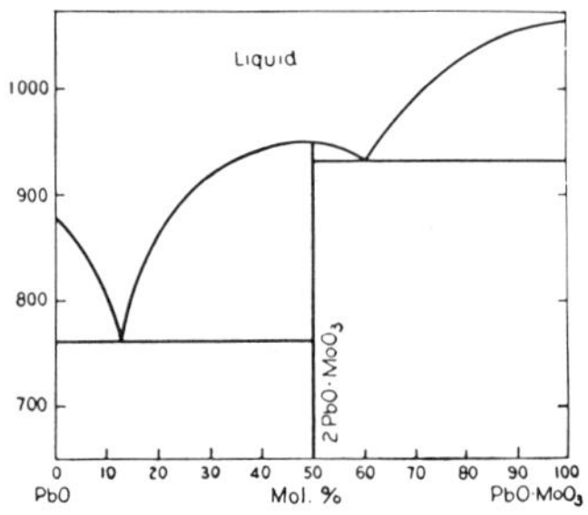

FIG. 292

PbO–MoO₃

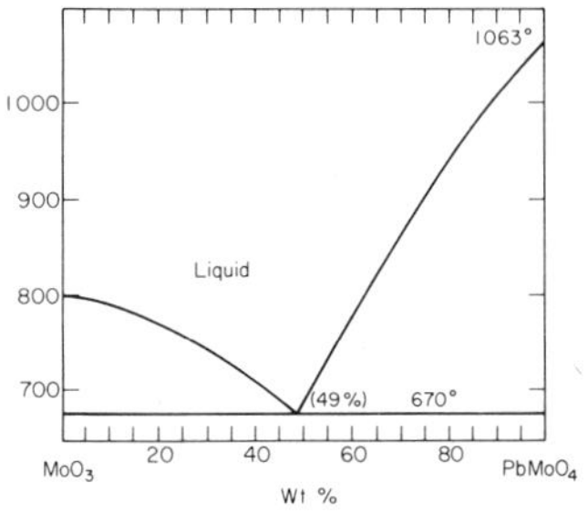

FIG. 4355

Bi₂O₃–RO

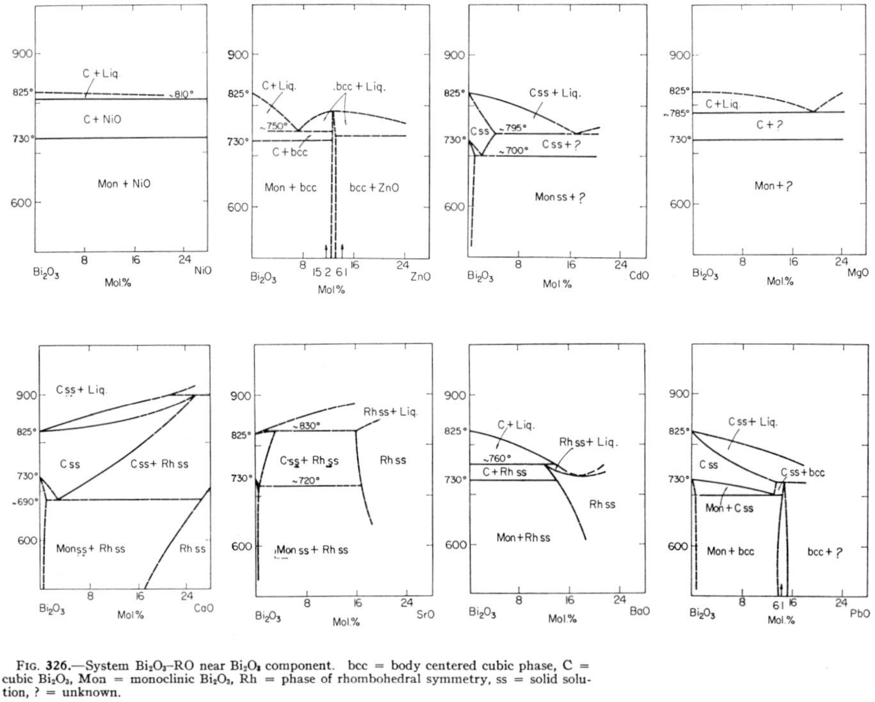

FIG. 326.—System Bi₂O₃–RO near Bi₂O₃ component. bcc = body centered cubic phase, C = cubic Bi₂O₃, Mon = monoclinic Bi₂O₃, Rh = phase of rhombohedral symmetry, ss = solid solution, ? = unknown.

PbO–Al$_2$O$_3$–SiO$_2$

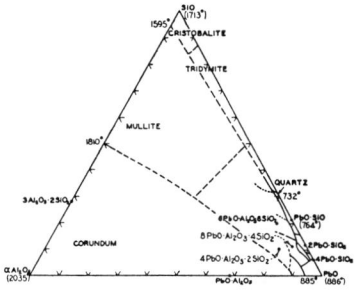

FIG. 737.— SiO at top vertex of diagram should read SiO$_2$.

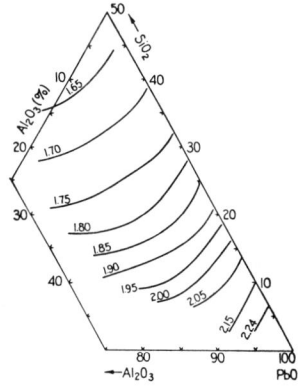

FIG. 739.—Indices of refraction of glasses in system PbO–Al$_2$O$_3$–SiO$_3$.

BaO–PbO–SiO$_2$

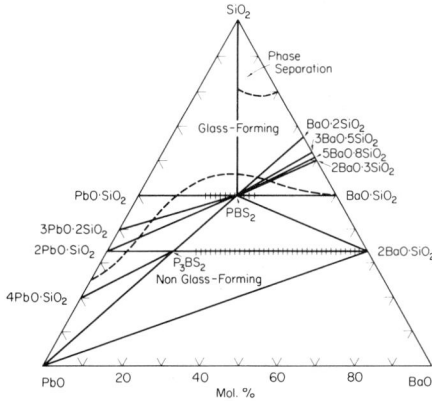

FIG. 2460.—System BaO–PbO–SiO$_2$ showing compatibility triangles. PBS$_2$ = PbO·BaO·2SiO$_2$ and P$_3$BS$_2$ = 3PbO·BaO·2SiO$_2$.

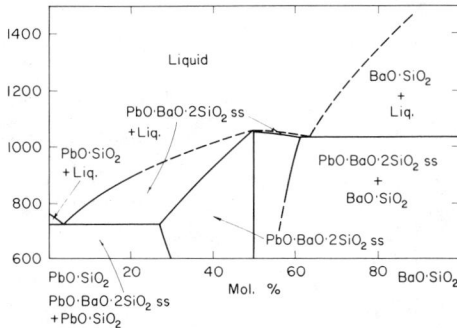

FIG. 2461

BaO–PbO–SiO₂ (concl.)

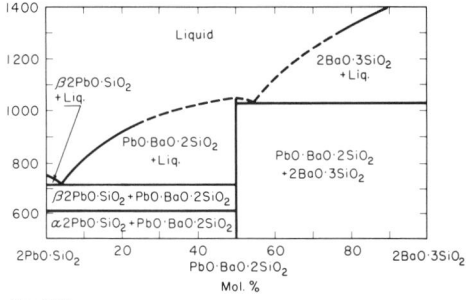

Fig. 2462

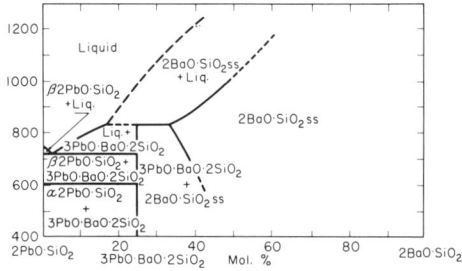

Fig. 2463

PbO–B₂O₃–SiO₂

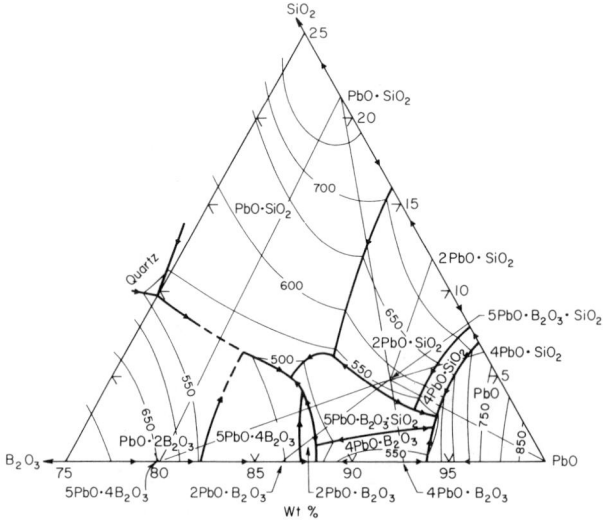

Fig. **4584.**—System B₂O₃–PbO–SiO₂, high-Pb region.

K$_2$O–PbO–SiO$_2$

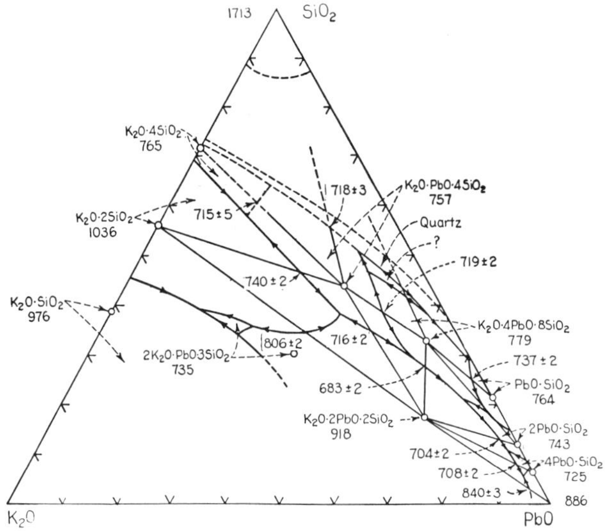

FIG. 404

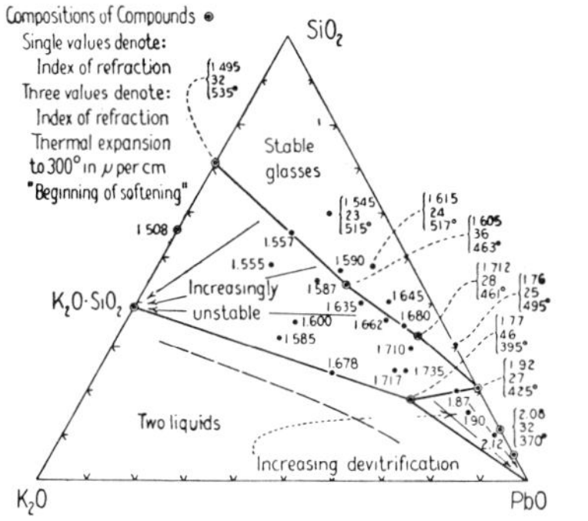

FIG. 405.—Summary of data on glasses of system K$_2$O–PbO–SiO$_2$.

MgO–PbO–SiO₂

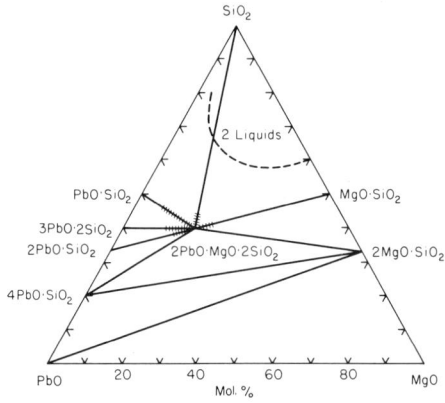

FIG. 2513.—System MgO–PbO–SiO₂; compatibility triangles. Hatched lines indicate solid solution.

Na₂O–PbO–SiO₂

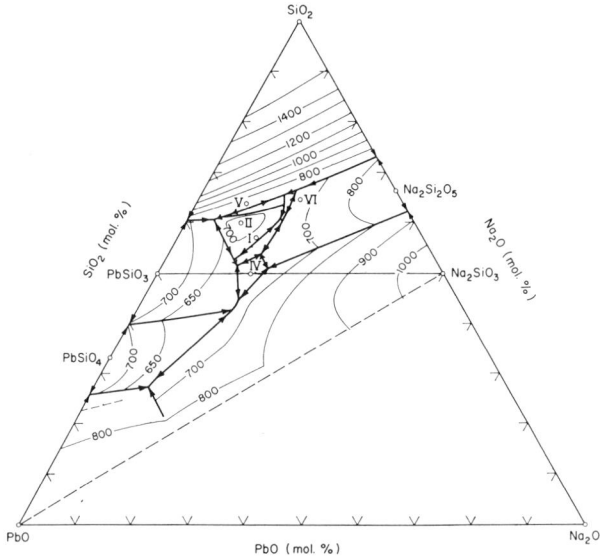

FIG. 494.—System Na₂SiO₃–PbO–SiO₂. I—Na₂O·2PbO·4SiO₂; II—Na₂O·3PbO·6SiO₂; IV—Na₂O·2PbO·3SiO₂; V—Na₂O·3PbO·7SiO₂; VI—3Na₂O·3PbO·11SiO₂.

On the PbO–SiO₂ boundary the compound labeled PbSiO₄ should read Pb₂SiO₄.

PbO–B₂O₃–TiO₂

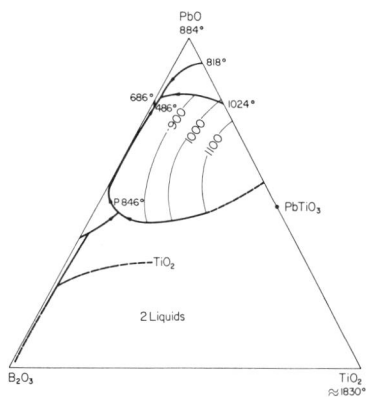

Fig. 743

PbO–Bi₂O₃–MoO₃

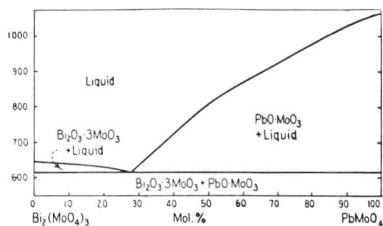

Fig. 744

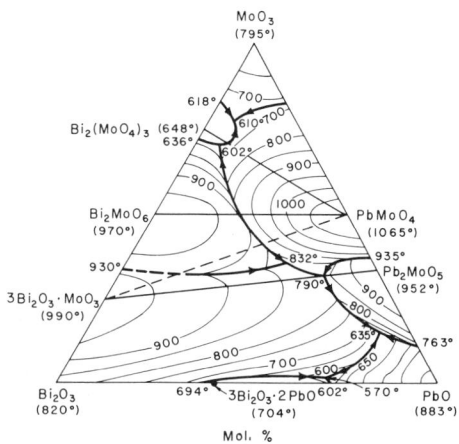

Fig. 2552

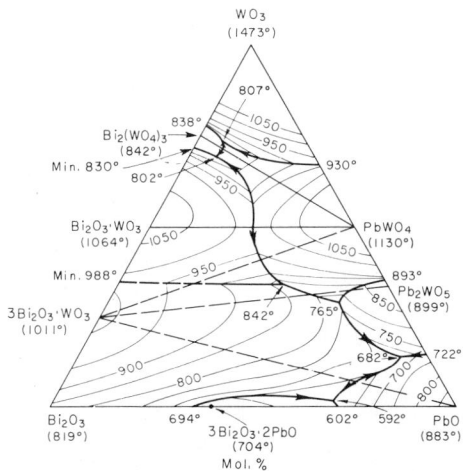

Fig. 2553

PbO–SiO₂–P₂O₅

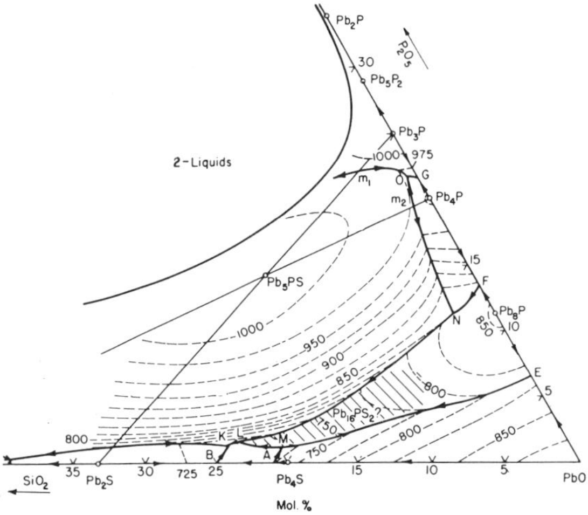

PbO–SiO$_2$–P$_2$O$_5$

FIG. 746

PbO–TiO₂–ZrO₂

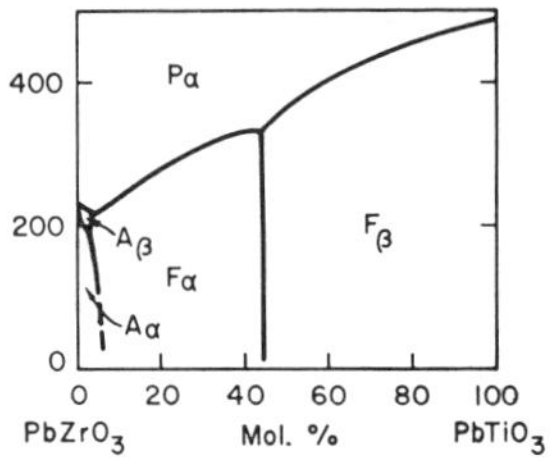

PbO–TiO$_2$–ZrO$_2$

Fig. 747.—System PbTiO$_3$.–PbZrO$_3$. P = paraelectric, cubic phase; Aα = antiferroelectric; orthorhombic phase; Aβ = antiferroelectric; Fα = ferroelectric, rhombohedral phase; Fβ = ferroelectric, tetragonal phase.

PbO–TiO₂–V₂O₅

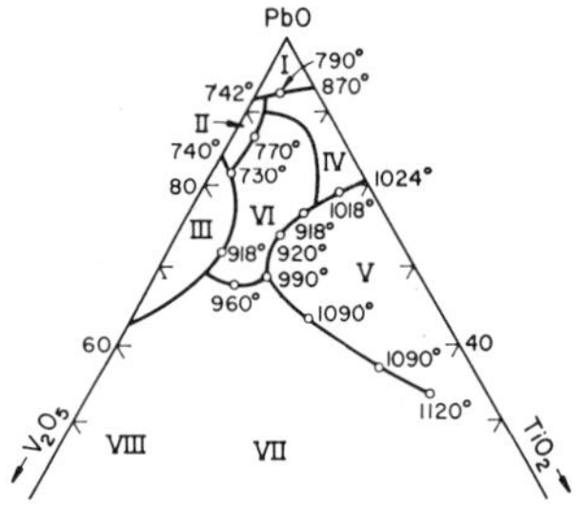

PbO–TiO$_2$–V$_2$O$_5$

FIG. 748.—System PbO–TiO$_2$–V$_2$O$_5$. I, PbO; II. 8PbO·V$_2$O$_5$; III, 3PbO·V$_2$O$_5$; IV, 2PbO·TiO$_2$; V, PbO·TiO$_2$; VI, 10PbO·V$_2$O$_5$·TiO$_2$; VII, TiO$_2$; VIII, glassy-looking phase.

PbO–TiO₂–ZrO₂

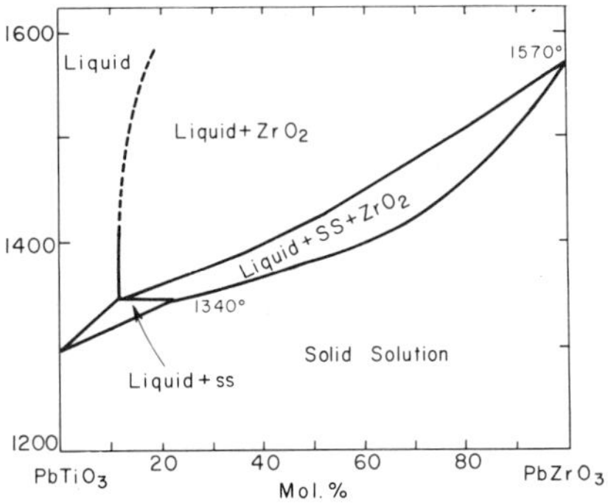

PbO–TiO$_2$–ZrO$_2$

FIG. 2560

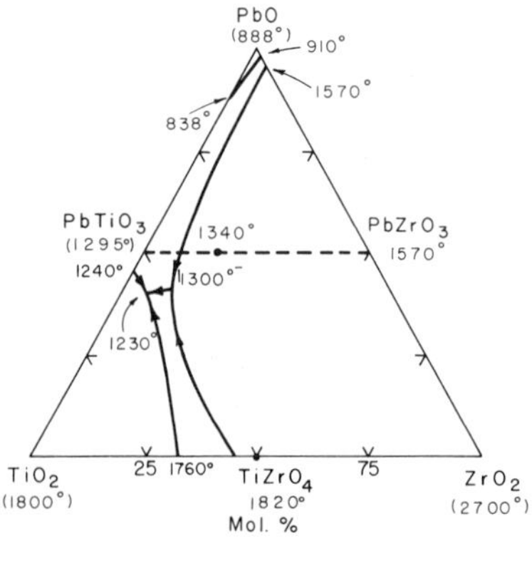

FIG. 2561

PbF$_2$-PbO-B$_2$O$_3$-Fe$_2$O$_3$-Y$_2$O$_3$ **KF-PbO-Nd$_2$O$_3$-ZrO$_2$-Ta$_2$O$_5$**

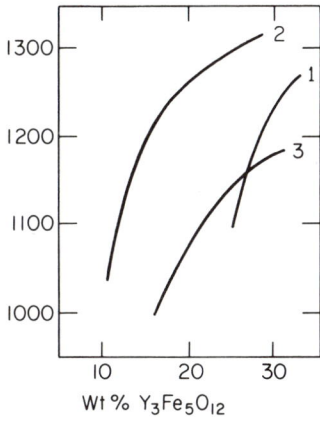

Fig. **6058**—System PbO-B$_2$O$_3$-Y$_2$O$_3$-Fe$_2$O$_3$-PbF$_2$. Solubility of Y$_3$Fe$_5$O$_{12}$ in three melts

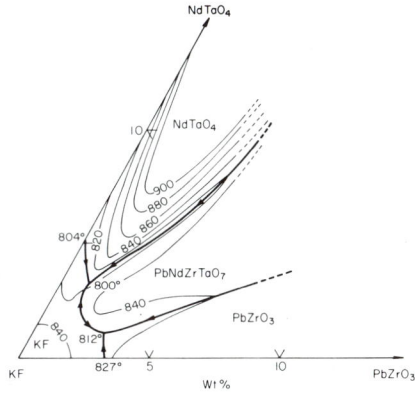

Fig. **6057**